NEW TRENDS IN ION EXCHANGE STUDIES

Edited by **Selcan Karakuş**

New Trends in Ion Exchange Studies

http://dx.doi.org/10.5772/intechopen.72067
Edited by Selcan Karakuş

Contributors

Arbakariya Ariff, Fadzlie Wong Faizal Wong, Murni HaliDove823tan319, Rafael Ribadeneira, Sergio Castañeda-Ramírez, Rajendra Sukhadeorao Dongre, José Ramón, Alejandro Manzano-Ramírez, Eric M. Rivera-Muñoz, Rodrigo Velázquez-Castillo, Miguel Apátiga-Castro, Rufino Nava, Aarón Rodríguez-López, Selcan Karakus

Notice

First published in London, United Kingdom, 2018 by IntechOpen
IntechOpen is the global imprint of INTECHOPEN LIMITED, registered in England and Wales, registration number: 11086078, The Shard, 25th floor, 32 London Bridge Street
London, SE19SG – United Kingdom
Printed in Croatia

British Library Cataloguing-in-Publication Data
A catalogue record for this book is available from the British Library

Additional hard copies can be obtained from orders@intechopen.com

New Trends in Ion Exchange Studies, Edited by Selcan Karakuş
p. cm.
Print ISBN 978-1-78984-247-0
Online ISBN 978-1-78984-248-7

For EU product safety concerns:
IN TECH d.o.o., Prolaz Marije Krucifikse Kozulić 3, 51000 Rijeka, Croatia,
info@intechopen.com or visit our website at intechopen.com.

Meet the editor

Assistant Professor Selcan Karakuş is currently working in the Department of Chemistry, Istanbul University-Cerrahpasa (IU-C), Turkey. She received her Master of Science degree in Physical Chemistry from IU in 2006. She received her Doctor of Philosophy degree in Physical Chemistry from IU in 2011. She worked as a visiting researcher at the University of Massachusetts, Department of Polymer Science and Engineering. She has research experience in drug carrier systems, nanoparticles, nanocomposites, nanoemulsion self-assembled polymeric nanostructures and copolymer blends. She has worked on different projects funded by Istanbul University. She has published several research articles and a book chapter in this area.

Contents

Preface

This book covers new systems in technology that have developed our knowledge of ion exchange. This book discusses ion exchange resins to enhance cell growth; anion exchange membrane; nanosystems in ion exchange and ion exchange in environmental applications. The ion exchange system is used in bionanotechnology, cosmetic industry and water treatment. In this book, readers will read recent studies, applications and new technological developments on major properties of ion exchange systems. We sincerely thank our authors who have contributed with experience and knowledge to this book. Especially, our thanks go to Prof Dr Ayben Kilislioglu and the editorial team from IntechOpen Publishing for their assistance in preparing this book.

Selcan Karakuş
Istanbul University-Cerrahpasa
Istanbul, Turkey

Chapter 1

Comparative Antibacterial Effects of a Novel Copper and Silver-Based Core/Shell Nanostructure by Sonochemical Method

Selcan Karakus, Ezgi Tan, Merve Ilgar,
Ismail Sıtkı Basdemir and Ayben Kilislioglu

Additional information is available at the end of the chapter

http://dx.doi.org/10.5772/intechopen.81588

Abstract

In this study, the antibacterial effect of novel copper (Cu) and silver (Ag) metal-based core-shell nanostructures against *Escherichia coli* (*E. coli*-Gram negative) was investigated. The novel copper- and silver-based nanostructures were prepared separately by using nontoxic, biodegradable, and biocompatible biopolymers chitosan and guar gum-polyvinyl alcohol (GG-PVA), which were modified by inorganic phases SiO_2 and sepiolite. On the other hand, guar gum-PVA (GG-PVA) was modified by sepiolite, and this nanostructure was prepared only for silver. Besides, Cu was dispersed in a different biopolymer chitosan by sonochemical method in the presence and absence of SiO_2. X-ray photoelectron spectroscopy (XPS), scanning electron microscopy (SEM), and X-ray diffraction (XRD) techniques were used to characterize the surface chemistry and morphology of the core/shell nanostructure. Nanoscale zero-valent Cu (NZVCu) was found under thin CuO film according to the XPS results. SEM images showed that spherical Cu/CuO@SiO_2 nanostructures (~100 nm) were homogenously dispersed in the chitosan by using sonochemical method. Antibacterial property of the core-shell nanostructures was analyzed by well-diffusion method against *Escherichia coli* (*E. coli*-Gram negative). Cu/CuO@SiO_2 nanostructures were found very effective against the *E. coli* due to high ratio of NZVCu in the nanostructure.

Keywords: sonochemistry, chitosan, silica, core-shell nanostructure, guar gum

IntechOpen

1. Introduction

Nanostructures are novel materials obtained by dispersing a small amount of nano-sized filler into a biopolymer matrix while preserving the material biodegradability and nontoxicity [1]. Depending on the dispersion and size of the inorganic filler component in nanostructures, they can exhibit improved mechanical, physical, chemical, and barrier properties, and biological reactivity in comparison to pure biopolymers. Because of their nanometer-size dispersion, biopolymer-clay nanocomposites exhibit large-scale advances in the mechanical and physical properties compared with pure biopolymers [2]. Generally, nanoparticles are prepared by chemical reduction, co-precipitation, sol-gel method, hydrothermal synthesis, thermal reduction, microwave process, vacuum vapor deposition, and sonochemical method [3–6].

The metal core-shell nanoparticle has many significant usages such as particle separation, drug delivery, magnetic resonance imaging, Raman imaging, and biosensor applications [7]. Various multicomponent heterostructured metallic nanoparticles are widely used as antibacterial agents (zinc oxide (ZnO), silver (Ag), Ag@ SiO_2, iron (Fe), molybdenum oxide (MoO_3), cerium oxide nanoparticles (CeO_2), gold-silver (Au-Ag), zirconium oxide ZrO_2, aluminum oxide (Al_2O_3), and magnesium oxide (MgO)) [8–17]. Rai et al. showed that nanomaterial load, type of substance, size and shape of nanoparticles, surface functional groups, crystallinity concentration are significant factors for their antibacterial effects. In spite of the advantages, their toxicity and safety are barriers that limit their efficient and safe use [18]. Dizaj et al. explained that there are two approaches that explain the antibacterial effect of metal nanoparticles: (a) free metal ion and (b) oxidative stress [19].

The copper-based nanoparticles are preferred to gold or silver nanoparticles because of low cost; high surface area; good thermal, mechanical stability, antimicrobial activity, and UV-light barrier property; high-performance conductive material in various applications; and use in photovoltaic and photocatalytic fields owing to their narrow band gap (1.2 eV) [20–25].

Chakraborty et al. evidenced that the antibacterial role of the Cu(II) oxide nanoparticle was a "particle-specific effect" which caused cellular DNA damage through phospho-di-ester bond breakage [26]. Gomes et al. demonstrated that Cu-salts were more toxic than Cu-NPs because of the oxidative stress and differential gene expression [27]. Wahid et al. underlined that copper-based nanostructures are easily released out of human body and can be easily mixed with polymers [28]. Gotzmann et al. pointed out that *Bacillus subtilis* is more sensitive to copper, but *E. coli* and *Staphylococcus aureus* are more sensitive to silver and discovered that the silver/copper blend displayed interdependent forceful antibacterial property [29]. Lv et al. provided antibacterial Cu nanoparticle (dosage of 100 g/mL) for disinfection of drinking water. They explained that cell of *Escherichia coli* were killed because of the reactive oxygen species with H_2O_2 playing a key role [30].

Chitosan is a natural aminopolysaccharide, nontoxic, biocompatible, biodegradable, derived by the deacetylation of chitin, and widely preferred in the biocomposite material preparation processes [31–34]. It has antibacterial property in an acidic solution depending on the kind of chitosan, molecular weight, and the degree of polymerization [35, 36]. Tamayo et al. reported that the -OH and $-NH_2$ groups of chitosan can react with H^+ ions to generate protonized chitosan with -NH^{3+} functional groups in acetic acid medium. The size and shape of nanoparticles are associated with protonized -NH_3^+ chitosan on surfaces and it decreases the amount of agglomeration, so we can synthesize more stable nanostructure [37, 38]. In this study, we determined that the most important condition is to control the size of nanoparticle which increases owing to self-agglomeration and can be intercepted by the addition of chitosan. The chemical bonds (Si-O-Si bonds) on the silica surface are unsaturated because the surface is active during synthesis. When silica functionalized with chitosan, the NH_2-functional shell was covalently bounded to the surfaces of silica and the reducing ability of SiO_2 increased when functionalized with amino groups [39, 40].

Our hypothesis in this study is acoustic cavitation, and critical amount of the inorganic phase played a key role in synthesizing Chi/Cu/CuO@SiO_2 core-shell nanostructure via sonochemistry. These cavitations live through couple of cycles of the ultrasound medium and collapse in a few nanoseconds because of high-temperature (>1000 K) and pressure (>100 atm) conditions in the solution. The chemical bonds (Si-O-Si bonds) on the silica surface are unsaturated because the surface is reactive during the synthesis. When silica functionalized with chitosan, the NH_2-functional shell was attached via H-bonds to the surface of silica and the reducing ability of SiO_2 increased when functionalized with amino groups. SiO_2 was very effective in reduction of Cu^{2+} into elemental copper. We prepared nanostructures with unique antibacterial properties to destroy pathogenic microorganisms by using sonochemical method. Our aim was the synthesis of the novel core-shell nanoparticle which affects the rate of microbial growth at very low concentrations and produces free radicals (superoxide ($^{\cdot}O^{2-}$), hydroxyl radical ($^{\cdot}OH$), hydrogen peroxide (H_2O_2)) [41–45]. The well-diffusion method was used to determine the antibacterial activity against pathogen bacteria such as *Escherichia coli*. The growth inhibition zones appeared in the agar layer. We found that the mechanism is founded on self-assembly reduction/oxidation reactions that occur among copper, SiO_2, and chitosan.

Silver is commonly used as a well-known antibacterial additive in polymer blends for food packaging and bionanotechnology applications. Guar gum is a biodegradable and eco-friendly biopolymer and is formed of mannose and galactose units [46]. Poly(vinyl alcohol) (PVA) is mixed with guar gum through hydrogen bonds and also has biocompatibility, biodegradability, and good mechanical properties [47]. In this study, Ag@Sepiolite-based nanostructure was encapsulated in a PVA/guar gum/matrix. Antibacterial property of the copper and silver core-shell nanostructures was analyzed by well-diffusion method against *Escherichia coli* (*E. coli*-Gram negative). Cu/CuO@SiO_2 nanostructures were found to be very effective against the *E. coli* due to high ratio of NZVCu in the nanostructure. According to this research, the major role of silica in Cu-based core-shell nanostructures under ultrasonic effect was highlighted.

2. Experimental

2.1. Materials

Chitosan (low molecular weight) and cetyltrimethylammonium bromide (CTAB) were purchased from Sigma-Aldrich. Glacial acetic acid, sodium hydroxide, $AgNO_3$, silica gel, $NaBH_4$, and Copper(II) sulfate pentahydrate ($CuSO_4.5H_2O$) were purchased from Merck. In the experimental setup, all chemicals and reagents were analytical grade and used without further purification.

2.2. Preparation of the Cu core-shell nanostructure

Chitosan aqueous solutions of low molecular weight, 0.02 g (100 mL), were prepared by dissolving chitosan powder in 5% (v/v) glacial acetic acid. The homogenous solution was obtained by continuous mixing at 25°C. Solutions containing 0.01 M $CuSO_4.5H_2O$ and 0.02 M $NaBH_4$were prepared separately in a 10-mL deionized water and mixed with a magnetic stirrer drop by drop. Silica gel of weight 0.02 g was added. Then, they were sonicated for 15 minutes (20 kHz) with chitosan solutions under a nitrogen atmosphere by using a Bandelin SONOPULS homogenizer. The samples were dried at 45°C to a constant weight (**Figure 1**).

2.3. Preparation of the Ag core-shell nanostructure

Natural sepiolite of weight 5 g was stirred using 500-mL distilled water to remove the soluble impurities for 1 night. Cetyltrimethylammonium bromide (CTAB) of weight 2 g in 100 mL hot distilled water was added into sepiolite solution and was mixed for 24 hour. Then, the samples were dried overnight at 110°C to a constant weight.

Sepiolite/CTAB of weight 0.0025 g was added to each polymer solution, and they were sonicated for 5 minutes. Solution of 0.133 M, 2.5 mL $NaBH_4$ was added drop by drop to 0.133 M,

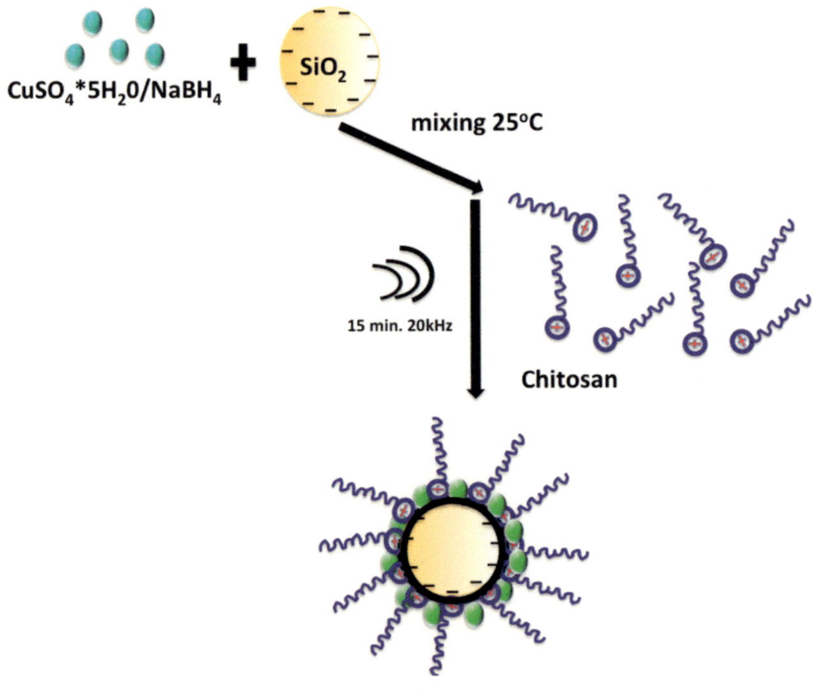

Figure 1. The mechanism of Chi/Cu/CuO@SiO_2 core-shell nanostructure.

2.5 mL $AgNO_3$ under nitrogen atmosphere while it was sonicated. Reduced Ag solution was added drop by drop to each PVA/Guar Gum polymer mixture (1:1) for 5 minutes under sonication under N_2 atmosphere. All samples were dried at 60°C. Also, the experiments were carried out without Ag and sepiolite/CTAB.

2.4. Characterization of core-shell nanostructure

2.4.1. X-ray diffraction (XRD) study

The diffraction patterns of the core/shell nanostructure were analyzed by XRD diffractometer (XRD, PAN analytical Xpert-Pro, Cu Ka: 1.5406 Å, a nickel monochromator filtering wave at 40 kV and 40 mA, with a 0.4/min at room temperature, Bruker D8 Advance X-ray Diffractometer).

2.4.2. Scanning electron microscopy (SEM)

The morphological studies were conducted using a Jeol/eo version 1.0 instrument Jsm6390 scanning electron microscope (SEM). The samples were coated with platinum before SEM analysis.

3. Results and discussion

3.1. XRD results

Figure 2 presents the XRD patterns of Chi/Cu/CuO@SiO_2, Chi/Cu/CuO, chitosan, and SiO_2. SiO_2 not only plays a key role in the reduction of copper salt into metallic copper during sonication-assisted mixing of inorganic and organic phases but also leads to the formation of a mineral. According to the XRD results, the diffraction data obtained were well matched with dioptase mineral (ICSD 100077 and PDF 33–487). The XRD pattern of nanostructure indicates the presence of Cu/CuO on the surface of SiO_2. A simple scheme was drawn to demonstrate the nanostructure (**Figure 3**). The XRD pattern of copper displayed characteristic peaks at 2 theta of 22.8, 24.3, 31.5, 34.0, 38.7, 48.9, 53.4, 58.0, 61.0, 66.1, 67.9, 72.2, and 74.9°, respectively [23].

Figure 4 presents the XRD patterns of (a) PVA-Guar Gum (b) PVA-Guar Gum-Sepiolite (c) PVA-Guar Gum Ag@Sepiolite core-shell nanoparticle. The XRD pattern of silver displayed characteristic peaks at 2 theta of 38.19, and 44.26°, respectively.

3.2. SEM results

According to the SEM images of the core-shell nanostructure, a homogenous structure was obtained by using sonication method. The SEM micrographs and EDAX mapping of Cu are shown in **Figures 5** and **6**.

3.3. X-ray photoelectron spectroscopy analysis

The high-resolution XPS spectra of the Chi/Cu/CuO, Chi/Cu/CuO@SiO_2 nanostructures for the copper regions are presented in **Figure 7**. The intensities of spectra of same elements can be compared with each other in both samples. For the Chi/Cu/CuO@SiO_2 sample, elemental Cu shows $2p_{3/2}$ peak positioned at 933.6 eV and $2p_{1/2}$ peak positioned at 953.4 eV with no satellite.

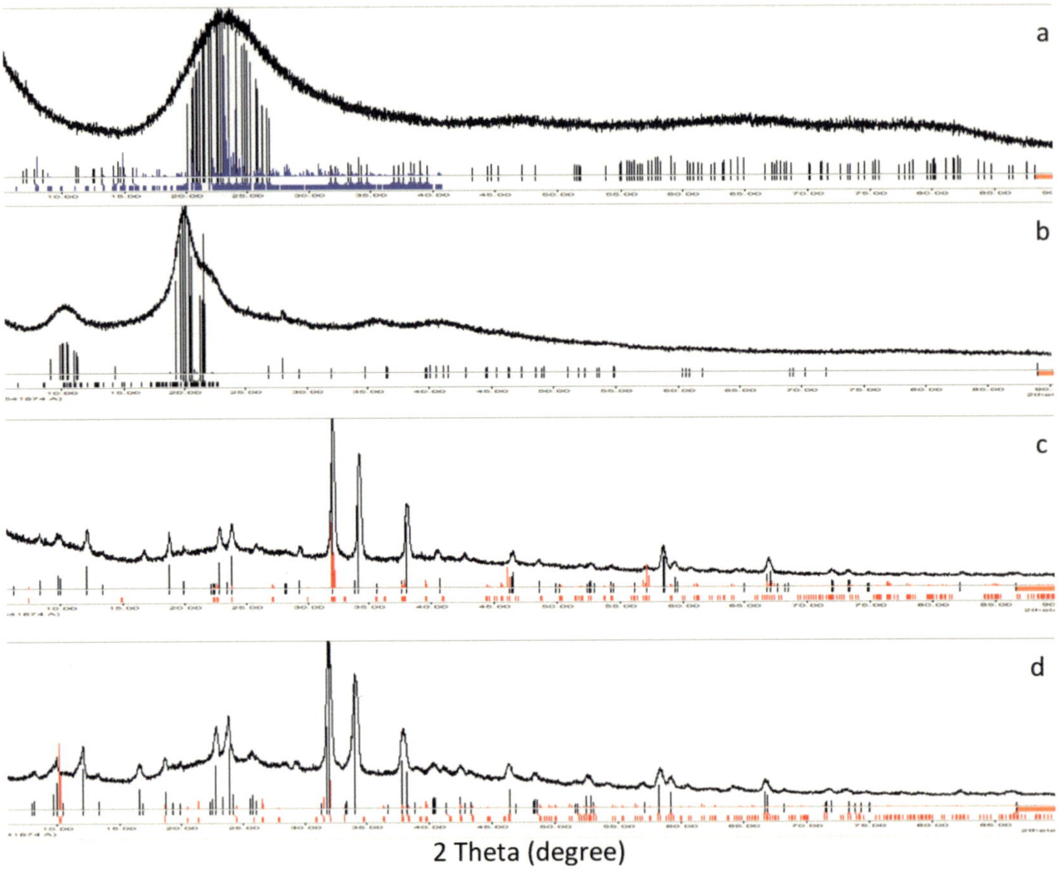

Figure 2. XRD patterns of (a) Chitosan (b) SiO_2 (c) Chi/Cu/CuO (d) Chi/Cu/CuO@SiO_2.

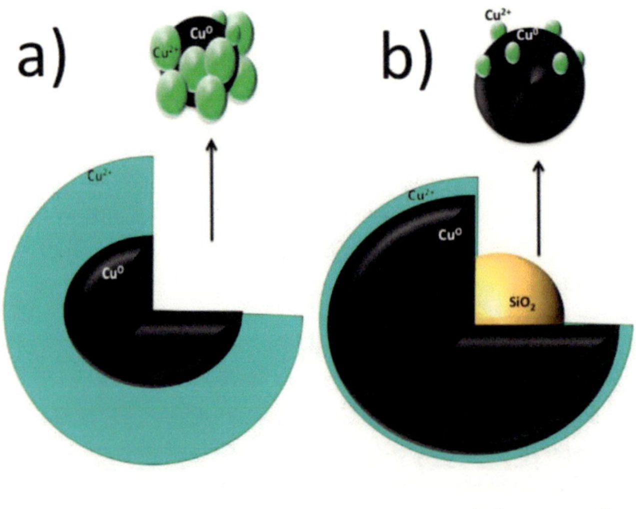

Figure 3. (a) Chi/Cu/CuO core-shell nanoparticle (b) Chi/Cu/CuO@SiO_2 core-shell nanoparticle.

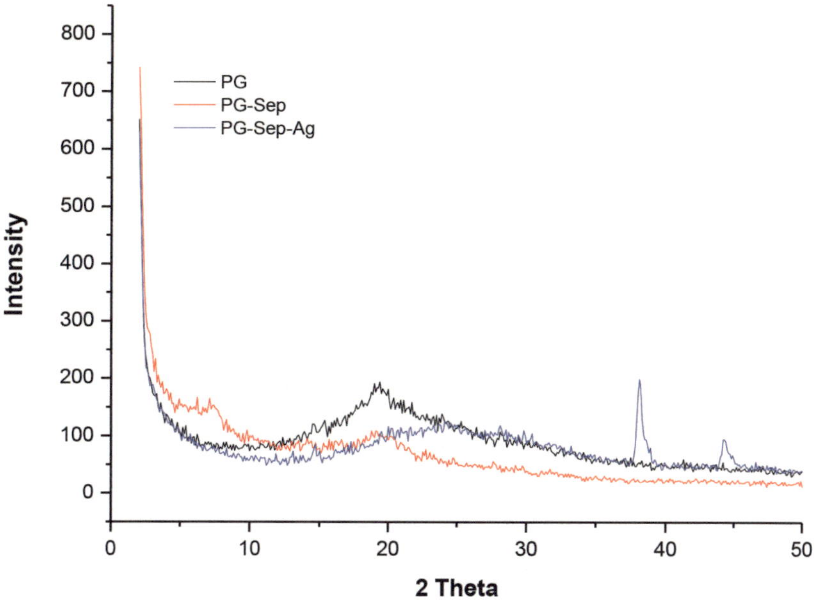

Figure 4. XRD patterns of (a) PVA-Guar Gum (b) PVA-Guar Gum-Sepiolite (c) PVA-Guar Gum Ag@Sepiolite core-shell nanoparticle.

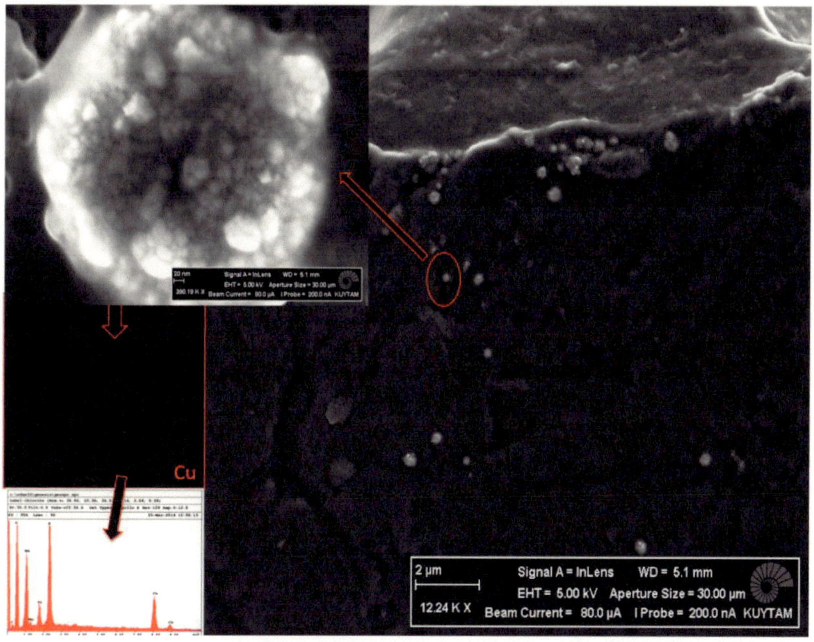

Figure 5. SEM images of the surface of $Chi/Cu/CuO@SiO_2$.

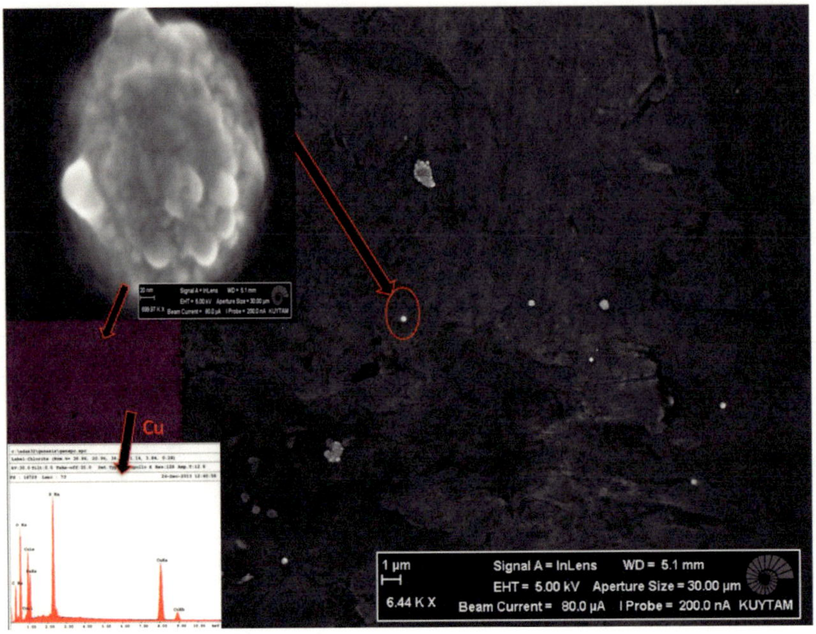

Figure 6. SEM images of the surface of Chi/Cu/CuO.

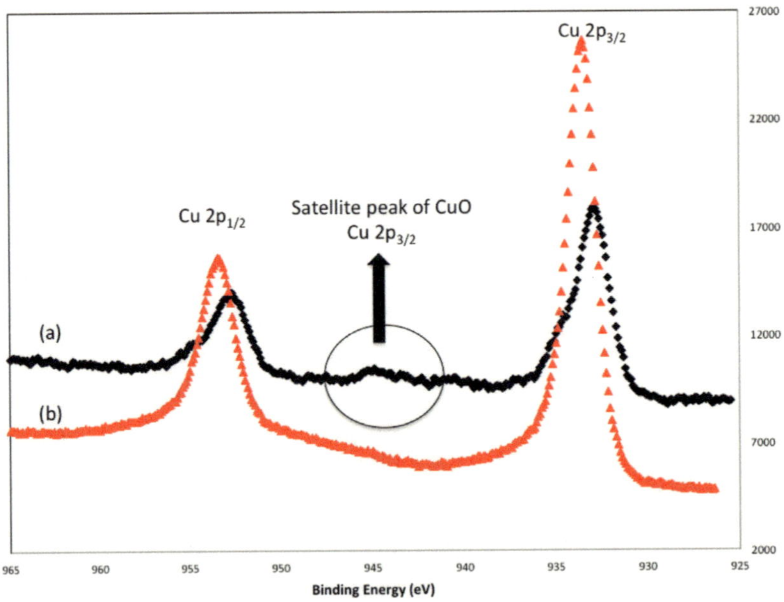

Figure 7. XPS pattern of (a) Chi/Cu/CuO (b) Chi/Cu/CuO@SiO_2.

For the Chi/Cu/CuO sample, the $2p_{3/2}$ maximum peak was positioned at 932.8 eV and $2p_{1/2}$ peak positioned at 952.5 eV. A satellite peak identifies the species as CuO due to surface oxidation of the ZVCu nanoparticles.

3.4. Antibacterial activity of the core-shell nanoparticle

The Gram-negative *Escherichia coli* bacteria (NCTC 10538) were cultured at 37°C on a shaking incubator and prepared by spreading of test organism on blood-enriched Mueller-Hinton agar, adjusting 10 mg of sample. The petri dishes were incubated to 35°C at 24 hours (180 rpm). The petri dishes were analyzed for the presence of a clear zone of inhibition. The antibacterial agents inhibited pathogenic bacteria growth leading to a clear, isolated zone in the petri dish. The power of antibacterial properties is proportional to the diameter of inhibition zone (disk diffusion). **Figure 8** shows a clear inhibition zone against *E. coli* microorganisms. The diameter of inhibition zone for nanoparticle against *E. coli* is approximately 20 mm, so that Chi/Cu/ $CuO@SiO_2$ has increased antibacterial performance against the bacteria. **Figure 9** shows that the diameter of inhibition zone for nanoparticle against *E. coli* is approximately 5 mm, so that PVA-Guar Gum Ag@Sepiolite has antibacterial property against the bacteria. Experimental results

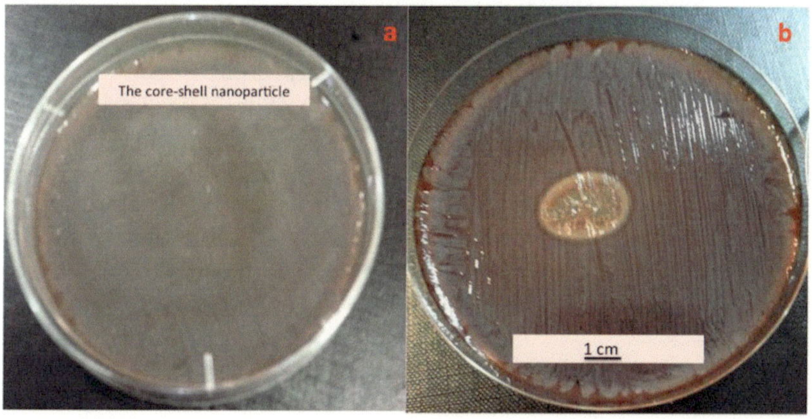

Figure 8. The zone of inhibition covered by Chi/Cu/$CuO@SiO_2$ nanoparticles against *E. coli*. (a) 1 hour; (b) 24 hours.

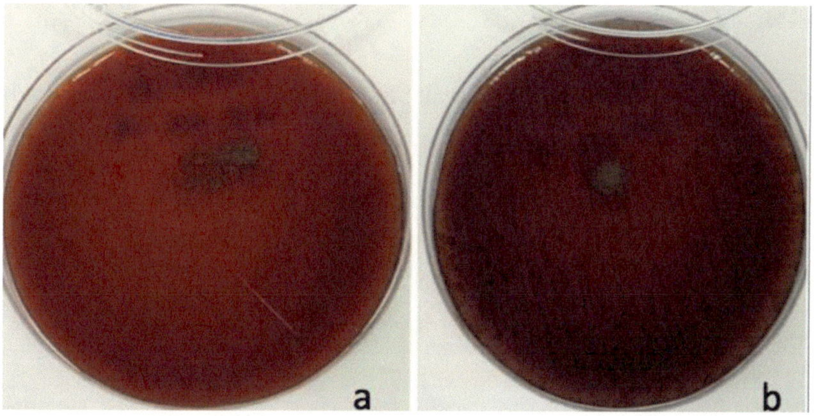

Figure 9. The zone of inhibition covered by PVA-Guar Gum Ag@sepiolite nanoparticles against *E. coli*. (a) 1 hour; (b) 24 hours.

showed that the size of nanoparticle, amount of nanoparticle, the oxidative stress, and particle-specific effect of SiO_2 are responsible factors for their improved antibacterial effect [45].

4. Conclusion

A novel antibacterial core/shell nanostructure was prepared by using chitosan which was modified by SiO_2 and Cu. Cu was dispersed in chitosan by sonochemical method in the presence and absence of SiO_2. When sonication was applied during preparation, the critical amount of SiO_2 and Cu caused the formation of dioptase mineral. SiO_2 led to the reduction in copper salt into metallic copper during dispersion by sonication and also produced a dioptase mineral.

The specific advantages of the preparation of this novel Cu/CuO@SiO_2 core/shell nanostructure included: (i) a low-cost, simple, convenient, and easily feasible sonochemical method, (ii) high content of Cu(0) nanoparticle in the targeting antibacterial agent, (iii) the novel nanoparticle has high bactericidal capacity, and (iv) critical mass production of the antibacterial core-shell nanoparticle. According to the results of the analysis, one could say that the copper-based nanostructure was found more effective against *E. coli* than the silver-based nanostructure due to the major role of silica in Cu-based core-shell nanostructures by using sonochemical method.

Acknowledgements

This work was supported by the research fund of Istanbul University project number 13145. The authors acknowledge the support on SEM and XPS analysis from Koc University Surface Science and Technology Center (KUYTAM).

Author details

Selcan Karakus*, Ezgi Tan, Merve Ilgar, Ismail Sıtkı Basdemir and Ayben Kilislioglu

*Address all correspondence to: selcan@istanbul.edu.tr

Department of Chemistry, Faculty of Engineering, Istanbul University-Cerrahpasa, Istanbul, Turkey

References

[1] Bordes P, Pollet E, Avérous L. Nano-biocomposites: Biodegradable polyester/nanoclay systems. Progress in Polymer Science. 2009;**34**:125-155

[2] Ke T, Le Y, Wang JX, Chu GW, Chen JF, Shao L. Cu nanoparticle preparation in a tube-in-tube microchannel reactor and encapsulation by silica. Materials Letters. 2010;**64**: 1717-1719

[3] Chowdhury MNK, Beg MDH, Khan MR, Mina MF. Synthesis of copper nanoparticles and their antimicrobial performances in natural fibres. Materials Letters. 2013;**98**:26-29

[4] Im HJ, Jung EC. Colloidal nanoparticles produced from Cu metal in water by laser ablation and their agglomeration. Radiation Physics and Chemistry. 2016;**118**:6-10

[5] Antonoglou O, Giannousi K, Arvanitidis J, Mourdikoudis S, Pantazaki A, Dendrinou-Samara C. Elucidation of one step synthesis of PEGylated CuFe bimetallic nanoparticles. Antimicrobial activity of CuFe@PEG vs Cu@PEG. Journal of Inorganic Biochemistry. 2017;**177**: 159-170

[6] Salavati-Niasari M, Hosseinzadeh G, Davar F. Synthesis of lanthanum hydroxide and lanthanum oxide nanoparticles by sonochemical method. Journal of Alloys and Compounds. 2011;**509**:4098-4103

[7] Chen Y, Wang Z, Chen X, Zeng D, Li M, Peng DL. Solution preparation of alloy core-shell nanoparticles: The case of Ni-Cu@Au-Cu nanoparticles. Materials Letters. 2013;**99**:180-183

[8] Joe A, Park SH, Shim KD, Kim DJ, Jhee KH, Lee HW, et al. Antibacterial mechanism of ZnO nanoparticles under dark conditions. Journal of Industrial and Engineering Chemistry. 2017;**45**:430-439

[9] Alimunnisa J, Ravichandran K, Meena KS. Synthesis and characterization of Ag@SiO_2 core-shell nanoparticles for antibacterial and environmental applications. Journal of Molecular Liquids. 2017;**231**:281-287

[10] Asghara MA, Zahir E, Shahid SM, Khan MN, Asghar MA, Iqbal J, et al. Iron, copper and silver nanoparticles: Green synthesis using green and black tea leaves extracts and evaluation of antibacterial, antifungal and aflatoxin B1 adsorption activity. LWT - Food Science and Technology. 2018;**90**:98-107

[11] Najafi M, Abbasi A, Masteri-Farahani M, Janczak J. Sonochemical preparation of bimetallic (Cu/Mo) oxide nanoparticles as catalysts for dye degradation under mild conditions. Polyhedron. 2015;**93**:76-83

[12] Gopinathan E, Viruthagiri G, Shanmugam N, Sathiya Priya S. Optical, surface analysis and antibacterial activity of ZnO-CuO doped cerium oxide nanoparticles. Optik. 2015;**126**: 5830-5835

[13] Yu X, Li J, Shi T, Cheng C, Liao G, Fan J, et al. A green approach of synthesizing of Cu-Ag core-shell nanoparticles and their sintering behavior for printed electronics. Journal of Alloys and Compounds. 2017;**724**:365-372

[14] Banerjee M, Sharma S, Chattopadhyay A, Ghosh SS. Enhanced antibacterial activity of bimetallic gold-silver core-shell nanoparticles at low silver concentration. Nanoscale. 2011;**3**:5120-5125

[15] Zhang Y, Wang L, Xu Y. ZrO_2 solid superacid porous shell void TiO_2 coreparticles (ZVT) polyvinylidene fluoride (PVDF) composite membranes with anti-fouling performance for sewage treatment. Journal of Membrane Science. 2015;**260**:258-268

[16] Ansari MA, Khan HM, Khan AA, Cameotra SS, Saquib Q, Musarrat J. Interaction of Al_2O_3 nanoparticles with *Escherichia coli* and their cell envelope biomolecules. Journal of Applied Microbiology. 2014;**116**:772-783

[17] Tang ZX, Lv BF. MgO nanoparticles as antibacterial agent: Preparation and activity. Brazilian Journal of Chemical Engineering. 2014;**31**:591-601

[18] Rai M, Ingle AP, Pandit R, Paralikar P, Gupta I, Chaud MV, et al. Broadening the spectrum of small-molecule anti bacterials by metallic nanoparticles to overcome microbial resistance. International Journal of Pharmaceutics. 2017;**532**:139-148

[19] Dizaj SM, Lotfipour F, Barzegar-Jalali M, Zarrintan MH, Adibkia K. Antimicrobial activity of the metals and metal oxide nanoparticles. Materials Science and Engineering: C. 2014; **44**:278-284

[20] Shankar S, Rhim JW. Effect of copper salts and reducing agents on characteristics and antimicrobial activity of copper nanoparticles. Materials Letters. 2014;**132**:307-311

[21] Ghasemi N, Jamali-Sheini F, Zekavati R. CuO and Ag/CuO nanoparticles: Biosynthesis and antibacterial properties. Materials Letters. 2017;**196**:78-82

[22] Li J, Shi T, Feng C, Liang Q, Yu X, Fan J, et al. The novel Cu nanoaggregates formed by 5 nm Cu nanoparticles with high sintering performance at low temperature. Materials Letters. 2018;**216**:20-23

[23] Oun AA, Rhim JW. Carrageenan-based hydrogels and films: Effect of ZnO and CuO nanoparticles on the physical, mechanical, and antimicrobial properties. Food Hydrocolloids. 2017;**67**:45-53

[24] Gutiérreza MF, Malaquiasa P, Matosa TP, Szesza A, Souzaa S, Bermudeza J, et al. Mechanical and microbiological properties and drug release modeling of an etch-and-rinse adhesive containing copper nanoparticles. Dental Materials. 2017;**33**:309-320

[25] Reddy KR. Green synthesis, morphological and optical studies of CuO nanoparticles. Journal of Molecular Structure. 2017;**1150**:553-557

[26] Chakraborty R, Sarkar RK, Chatterjee AK, Manju U, Chattopadhyay AP, Basu T. A simple, fast and cost-effective method of synthesis of cupric oxide nanoparticle with promising antibacterial potency: Unraveling the biological and chemical modes of action. Biochimica et Biophysica Acta. 2015;**1850**:845-856

[27] Gomes SIL, Murphy M, Nielsen MT, Kristiansen SM, Amorim MJB, Scott-Fordsmand JJ. Cu-nanoparticles ecotoxicity—Explored and explained? Chemosphere. 2015;**139**:240-245

[28] Wahid F, Wang HS, Lu YS, Zhong C, Chu LQ. Preparation, characterization and antibacterial applications ofcarboxymethyl chitosan/CuO nanocomposite hydrogels. International Journal of Biological Macromolecules. 2017;**101**:690-695

[29] Gotzmann G, Jorsch C, Wetzel C, Funk HWR. Antimicrobial effects and dissolution properties of silver copper mixed layers. Surface and Coating Technology. 2018;**336**: 22-28, article in press. DOI: 10.1016/j.surfcoat.2017.09.036

[30] Lv Q, Zhang B, Xing X, Zhao Y, Cai R, Wang W, et al. Biosynthesis of copper nanoparticles using Shewanella loihica PV-4with antibacterial activity: Novel approach and mechanisms investigation. Journal of Hazardous Materials. 2018;**347**:141-149

[31] Pillai CKS, Paul W, Sharma CP. Chitin and chitosan polymers: Chemistry, solubility and fiber formation. Progress in Polymer Science. 2009;**34**:641-678

[32] Chandy T, Sharma CP. Chitosan—as a biomaterial. Biomaterials, Artificial Cells, and Artificial Organs. 1990;**18**:1-24

[33] Paul W, Sharma CP. Chitosan, a drug carrier for the 21st century: A review. STP Pharma Sciences. 2000;**10**:5-22

[34] Muzzarelli RAA, Muzzarelli C. Chitosan chemistry: Relevance to the biomedical sciences. Advances in Polymer Science. 2005;**186**:151-209

[35] Qi L, Xu Z, Jiang X, Hu C, Zou X. Preparation and antibacterial activity of chitosan nanoparticles. Carbohydrate Polymers. 2004;**339**:2693-2700

[36] Sotelo-Boyás ME, Correa-Pacheco ZN, Bautista-Baños S, Corona-Rangel ML. Physicochemical characterization of chitosan nanoparticles and nanocapsules incorporated with lime essential oil and their antibacterial activity against food-borne pathogens. LWT - Food Science and Technology. 2017;**77**:15-20

[37] Tamayo L, Azócar M, Kogan A, Riveros M, Páez M. Copper-polymer nanocomposites: An excellent and cost-effective biocide for use on antibacterial surfaces. Materials Science & Engineering C: Materials for Biological Applications. 2016;**69**:1391-1409

[38] Javed R, Ahmed M, ul Haq I, Nisa S, Zia M. PVP and PEG doped CuO nanoparticles are more biologically active: Antibacterial, antioxidant, antidiabetic and cytotoxic perspective. Materials Science and Engineering: C. 2017;**79**:108-115

[39] Guan Z, Shu Y, Maa Y, Wan J. Factors affecting the physicochemical properties of the modified core/shell NH_2-SiO_2@NZVI nanoparticles. Colloids and Surfaces A: Physicochemical and Engineering Aspects. 2015;**482**:18-26

[40] Markova-Deneva I. Infrared spectroscopy investigation of metallic nanoparticles based on copper, cobalt, and nickel synthesized through borohydride reduction method. Journal of Chemical Technology and Metallurgy. 2010;**45**(4):351-378

[41] Dolores R, Raquel S, Adianez GL. Sonochemical synthesis of iron oxide nanoparticles loaded with folate and cisplatin: Effect of ultrasonic frequency. Ultrasonics Sonochemistry. 2015;**23**:391-398

[42] Arakha M, Roy J, Nayak PS, Mallick B, Jha S. Zinc oxide nanoparticle energy band gap reduction triggers the oxidative stress resulting into autophagy-mediated apoptotic cell death. Free Radical Biology & Medicine. 2017;**110**:42-53

[43] Pandey S, Ramontja J. Sodium alginate stabilized silver nanoparticles—silica nanohybrid and their antibacterial. International Journal of Biological Macromolecules. 2016;**93**: 712-723

[44] Raut SS, Kamble SP, Kulkarni PS. Efficacy of zero-valent copper (Cu(0)) nanoparticles and reducing agents for dechlorination of mono chloroaromatics. Chemosphere. 2016;**159**: 359-366

[45] Chen S, Guo Y, Zhong H, Chen S, Li J, Ge Z, et al. Synergistic antibacterial mechanism and coating application of copper/titanium dioxide nanoparticle. Chemical Engineering Journal. 2014;**256**:238-246

[46] Mukherjee S, Mukhopadhyay S, Zafri MZB, Zhana X, Hashim MA, Gupta BS. Application of guar gum for the removal of dissolved lead from wastewater. Industrial Crops and Products. 2018;**111**:261-269

[47] Das T, Yeasmin S, Khatua S, Acharya K, Bandyopadhyay A. Influence of a blend of guar gum and poly(vinyl alcohol) on long term stability, and antibacterial and antioxidant efficacies of silver nanoparticles. RSC Advances. 2015;**5**:54059-54069

Chapter 2

Chitosan-Derived Synthetic Ion Exchangers: Characteristics and Applications

Rajendra Sukhadeorao Dongre

Additional information is available at the end of the chapter

http://dx.doi.org/10.5772/intechopen.78964

Abstract

Today growing science and technological needs explored various biopolymers to procure novel utilities in its modern developments. Consequently, polysaccharides embraced huge prospective and vastly caters such desired growing needs. Amid, chitin the second most ubiquitous after cellulose comprise of β-[1,4]-2-acetamido-2-deoxy-d-glucose flexible skeleton undergo alteration for requisite physico-chemical features and its highly sophisticated utility superseded counterpart cellulose. Chitosan have unique parameters namely bio-compatibility, non-toxicity, hemeostaticity, anti-microbials which offer competent solutions of many challenging problems. Thus, many products namely biomarkers, biosensors, quantum dots are fabricated via adoptable productive chitosan matrixes. Advancement in chitosan chemistry proffers unambiguous industrial utility in cosmetics, pharmaceuticals, nanobiotechnology, water purifications etc. Chitosan composites own enhanced muco-adhesivity that aids pharmacological safe and successful DNA/SiRNA/tissue releases with bioavailability at target specific carriers. ZnO, ZnS, TiO2 filled/imposed in chitosan and resultant hybrids, quantum dots, surface active microcapsules and nanoparticles are used as biosensors, bio-markers, adsorbents that proffers revolutionary medical usage. Nanointegrated chitosan own complementary strengths and possess assorted utility namely nano-electronic high-resolution devices, for in-vivo imaging, diseases diagnosis, generating new therapeutic and smart tissue engineering scaffolds. Novel modalities with innovative formulations are skillfully designed via chitosan matrix for myriad benefit in biology, chemistry, polymer, and pharmaceutics are displayed in this chapter.

Keywords: chitosan, chitin, biopolymers, biomarker, biosensor, drug delivery, water, membrane, nanotechnology, pharmaceuticals

1. Introduction

Bioactive molecules and natural polymers obtained from animals, plants, fungi and bacteria have ever fascinated the global scientists due to their survival profile for our environment and life [1]. Amid these bio-polymers, polysaccharides namely cellulose, chitin and starch pertaining functions in genetics as cell wall architecture, in biology as energy storage, in material science for bio-composite designing/discovery and for physic-chemical adsorption in chemistry respectively. Today S&T indispensably practices biopolymers for inclusive development in perspective of connectivity, communicative transactions and economic progression in medical sciences besides to cope up advanced biotechnology demands. Twenty-first century scientific innovations and technological throughput ingeniously boosts myriad fields like pharmaceutics, environment and nanotechnology besides uplift our living by enabling reachable services and numerous reliable products in health care. Nevertheless, the requisite new innovative materials are being fabricated from both, chitin and chitosan matrixes due to inherent characteristics uniqueness as futuristic, multi-functional; novel, versatile and peculiar diversity which is devoid in conventional/counterparts.

Chitin is the second most abundant biopolymers after cellulose as explored for R&D besides numerous remarkable papers/patents catering futuristic demands in S&T besides utility in biochemistry, organic/polymer chemistry, pharmacology and medicines. In fact chitin polysaccharide comprise of β-[1,4]-2-acetamido-2-deoxy-D-glucose repetitive units i.e., *N*-acetylglucosamine linked via glucosidic bonds that are facile for physico-chemical adaption to yield into new bio-composites, hybrids and blends owing superior capabilities widespread over cellulose. Deacetylated chitin called chitosan can be categorized by means of purity/quality found to possess exclusive features namely bio-compatibility, bio-degradable, non-toxicity, antimicrobials and hemeostatic. Advance nano-technology exploited superior, productive and widely derived matrixes from flexible chitin/chitosan, thus signifies their chemistry in science and technology [1]. About 1500 tons/year of chitin is usually produced throughout the world. In past decades, progressive bendable chitin/chitosan makes it an ideal model matrix for desired physic-chemical and/or enzymatic alterations/modulations as advantageous in modern S&T [1, 2]. Incredibly practical impact of chitosan in modern S&T is attributed to its liberal super-active $—NH_2$/—OH groups with subsequently performed diverse chemical modifications namely *N*-acylation/alkylation, *N*-quaternization and C-6 carboxylation. Literature reported that such substantial adaptation onto chitin/chitosan matrix persuades inclusive cationic charge via amine protonation to ammonium ion/NH_3^+ further improve its pH dependency (acid to alkaline) and resultant solubility [2]. Thus certain chitosan based materials owing wide putative applications are portrayed as below.

Chitosan matrixes extensively provides generous opportunities in biotechnology, theranostic and pharmaceuticals for drug/gene release, tissue engineering and wound healing respectively due to exceptional features like biodegradability, biocompatibility, antimicrobial profile, besides low toxicity, and immunogenicity [2–8]. This mini review signifies chitosan for advanced formulations in nano-science/biotechnology with reference to quantum dots, and nanoparticles, carbon dots, biosensors and biomarkers.

2. Advances in chitosan chemistry: unambiguous myriad applications in modern S&T

Chitosan own adaptable promising applications de-acetylated chitin derivative, despite its limited mechanical strength and solubility. The proactive functionalities of native chitosan like primary amino and hydroxy groups are facile to undergo many possible physical/chemical modifications/alternations through grafting and ionic interactions so as to yield assorted derivatives which can offer specific and requisite commercially utilities. The following sections comprise contemporary research in chitin/chitosan matrix towards applications in numerous industrial and clinical fields.

2.1. Chitin/chitosan processing

Chitin/chitosan are commercial natural resources, as well 150,000 metric tons as a sea-food industry wastes generate via processing from crab, shrimp, shell-fish, krill, clam, oyster and squid processing beside extracted from certain fungi. Compared to synthetics/counterparts, chitin/chitosan own exclusive nitrogen content and high concentrations of proteins along with calcium carbonate as procured via physic-chemical/biological extractions/treatments. Further chitin undergoes de-acetylation by means of alkali refluxing yields 85% product called chitosan. Chronological step-wise treatments like deproteinization, demineralization, decolouration and de-acetylations performed to produce chitosan from sea-food shown below:

Crustacean shell-size reduction—protein separation by NaOH/alkali reflux—washing—demineralization by HCl/mineral acid—dewatering—chitin decolouration—deacetylation by NaOH—washing/dewatering yield 85% chitosan.

2.2. Solubility of chitin/chitosan

Chitin/chitosan owes anhydro-glucoside link akin to cellulose, yet characteristics are vastly different as being strong hydrophobic insoluble in water/aqueous state. Many organic solvents, too cannot dissolve chitin/chitosan instead solvent mixtures are employed for its dissolution namely hexa-fluoroisopropanol, hexa-fluoroacetone, chloro-alcohol, 5% LiCl added dimethylacetamide, 30% aqueous acetic acid, aqueous *N*-methyl morpholine-*N*-oxide and aqueous mineral acids. Chitin and chitosan being highly basic (pH > 7) are soluble in various aqueous-organic/mineral-acid mixtures to form invariably poly-electrolytic jelly that's easily transformed into myriad compositions including salt, film, hybrid, chelate/complex and gels [2].

2.3. Chitin/chitosan salient treatments

Concentrated acid hydrolysis and drastic treatment onto chitin/chitosan produces serviceable β-D-glucosamine: amino sugar functional unit. Chitin/chitosan, based on deacetylation degree owns 6–8% w/v nitrogen/free $-NH_2$ being facile for varied chemical transformations/modifications namely *N*-acylation and Schiff-base reactions. Chitosan on treatment with keto-acids followed by sodium borohydride reduction yields proteic/non-proteic amino acid (1,3/1,6)-β-D-glucan and melanin linkage pertaining effective immuno-stimulant behaviour to boost immunity in human/animals. Chitosan undergoes many chemical reactions like forms *N*-acylation with acid anhydrides/acyl halides aldimine and ketimine with aldehyde and ketone respectively at NTP. Hydrogenation and *N*-alkylation with more bulky substituent weakens hydrogen bonding in chitosan skeleton; so *gets* swelled in water and retains film forming tendency in spite of hydrophobic alkyl chains. Further chitosan is quite versatile than chitin due to the presence of amino groups at the C-2 skeleton.

2.4. Chemical and biological properties of chitosan

Chitosan is linear polyamine skeleton owing free-active functionalities like primary amino and hydroxyl both can chelate transitional metals to form complexes. Biological chitosan is biocompatible natural polymer which is biodegradable to normal body constituents, hence safe and non-toxic to environment upon disposal. Chitosan binds to mammalian and microbial cells assertively own connective gum tissue regeneration and accelerate osteoblast formation as vital for bone regeneration, Moreover, chitosan possess board microbiological profile namely as hemostatic, fungistatic, spermicidal, antitumor, anticholesteremic besides acts as central nervous system depressant and immune adjuvant.

2.5. Chitosan derivatives

Chitin/chitosan is readily derivatized at primary amine, primary and secondary hydroxyl functionalities [3]. These derivatives own potential significance via further bi/poly-functional chemical reaction yields polymeric composites, blends, gels and industrially applied polyampholytes effective in remediation of pollution. Chitosan chelated complexes with transition metals acts as matrix for enzyme immobilizations [12]. Reactions with pure chitin have been carried out mostly in the solid state owing to the lack of solubility in ordinary solvents. Chitin/chitosan of about 50% degree of de-acetylation found to be water soluble [17] and used as feedstock for smooth modifications, through various solution phase transformation as enumerated below:

A. *N-phthaloylation of chitosan*: phthalic anhydride-DMF reacts with chitosan to yield *N*-phthaloylated chitosan which found to enhance solubility and affixed bulkiness due to breaking hydrogen of primary amine and averts hydrogen bonding to firm backbone/skeleton.

B. *Chitosan-sialic acid dendrond hybrid*: the water-solubility of chitosan matrix gets improved effectively via gallic acid and tri-ethylene glycol dendronized chitosan-sialic acid hybrid synthesis. Residual amine functionality can be further *N*-succinylated to impart great water solubility of such novel chitosan derivatives.

C. *Alkyl/aryl-thiocarbamoyl chitosan*: methyl/phenyl-thiocarbamoyl chitosan derivatives are prepared for selective metal ionic sorptions from aqueous solution/contaminated waters.

D. *Chitosan hydrogels*: chitosan hydrogels can be straight grafted by treatment with D,L-lactic and/or glycolic acid owing great interfacial water-chitosan interaction due to grafting with such acids. Chitosan side chains can physically cross-linked/aggregated to yield pH-sensitive hydrogels that owe potential biomedical functions namely wound dressing/healing, gene and drug delivery.

E. *Chitosan composite base quantum dots*: nano-cadmium sulphide doped chitosan improved thermal and mechanical besides aqueous solubility in parent bio-polymeric matrix. Efficient chitosan based CdS-QD/film/composite can be easily obtained by mixing chitosan, $Cd(Ac)_2$ and CdS in 1% aqueous acetic acid solution.

F. *Chitosan based nano-particals/composites*: gadolinium neutran capture therapy (Gd-NCT) induced chitosan-gadopentetic acid nanoparticles used for cancer therapy appropriate for intra-tumoral treatment. Flexible chitosan skeleton endure facile blends with synthetic polymers via mechano-chemical routes yields novel composites/nano-particles. Thermal performance and molecular signaling of synthetic polymeric matrix/composites are exclusively dissimilar from each/both components. These blends impart strong integrated chitosan cascades owing talented compatibility under mild conditions as utilized in smart and advanced drug-gene delivery.

2.6. Biological profile of chitin/chitosan

Human body fluids enzyme, lysozyme can dissolve some bacteria via cell-wall material cleaving, also facile to assemble in chitin/chitosan matrixes and impose myriad therapeutic usages in past decade. Certain vital medical usages of such chitosan matrixes including fibroplasias inhibition resulted wound healing/dressing, as absorbable sutures and supports in tissue/cell growth or differentiating tissues. Such chitosan based sutures found to resist attack in urine, bile and pancreatic juice while hasten to enhance wound healings dressings/textures that's hardly achieved by absorbable counterparts.

2.7. Industrial utility of chitin/chitosan matrixes

Apart from this chitin/chitosan apparel industrial usages and impending functions are exploited in wastewater/water purifications especially heavy metal and organic pollutant removal via chelation with wide scope and possibility. Chitosan base derivatives impart range of pharmaceutical and cosmetic products way from water treatment to plant/food protections.

2.7.1. Cosmetics

Unlike to other polyanionic hydrocolloids, this natural amino-polysaccharide chitosan encompassed hydrocolloids are exclusively cationic that gets viscous/semi-solid upon neutralization with aqueous organic acid solutions. Such viscous/semisolid chitosan facilitates intervene common integuments like skin covers and hairs. Chitin/chitosan both are fungicidal and fungi-static and thus compatible with other highly integrated biological counterparts used in such beauty products. Chitosan-alginate (1–10 μ size) microcapsules embodied various hydrophobic materials are utilized in range of cosmetic/beauty products. Substances absorbing harmful UV light and assorted dyes get linked to amino functionality of chitosan via covalent bonding. Consequently, chitosan enclosed compositions were developed by *Sonat* Company USA that can insert anti-oxidants, anti-allergic, and anti-inflammatory agents as novel depilatory designed in many areas of cosmetics like skin care, hair care, and curling hairs besides oral care. Chitin/chitosan and human hairs can harmonize mutually owing to opposite electrical charges as chitosan is cationic and hairs are anionic. Chitosan containing solution forms clear, elastic film or foam and create emulsifying action on hairs so, boots softness, smoothness, and mechanical strength. Chitosan based compositions also forms hydro-gel in aqueous alcohol thus used in shampoo, rinse, permanent wave agent and styling lotion besides hair spray/colorant and tonic. Several chitosan based compounds namely glyceryl chitosan, *n*-hydroxypropyl chitosan, quaternary hydroxypropyl chitosan, polyoxyalkylene chitosan, chitin sulphate and carboxymethyl chitosan own potential in hair care products. Chitosan derivatives are good candidate for skin care as they impart positive electrical charge, and due to high molecular weights it cannot infiltrate human skin, so act as a moisturizer and might compete with hyaluronic acid in this perspective. Chitin/chitosan both own wide utilities in formulation like creams, pack material, lotions, foundation, eye shadow, lipstick, cleansing, bath agents and nail enamel/lacquers. Rather diacid anhydride treated chitosan derivatives are best employed in many skin care products besides usage in paste, mouthwash and chew gum. Chitosan based salts are supplemented in toothpaste for mask silicon oxide unpleasant taste besides acts as powder binder to uphold granular shapes. Chitin materials are used as dental fillers that vitally absorb candida/thican, teeth sticking fungi, so recover cleaning false teeth.

2.7.2. *Chitosan based scaffolds in medical/clinical/pharmaceuticals*

Chitosan based skeletons are well-known in biomedical utility namely drug/gene release, wound dressing and advanced nanotechnology [1]. Chitosan based 2D/3D scaffolds like sponge, foam, gels fibers/film are more developed for modern tissue and bone engineering [2]. Notable *N*-trimethyl-*N*-octyl-chitosan scaffolds have established its potential utility as controlled drug delivery of hydroxyl-camptothecin anticancer agent [3]. Antitumor drug delivery is also reported with specially fabricated magnetic nano-iron doped maltosyl chitosan hybrids [2, 3]. Several clinical reports endorse impact chitosan matrixes in cancer chemoprevention due to precise nutrients encapsulation which imparts enhanced drug delivery in blood with fewer side effects on healthy cells [1, 2].

2.7.3. *Chitosan based matrixes in science and technology*

Synthetic plastic are produced @ 300 million tons/year with only 3% recyclability, rather 97% plastic wastes break down in oceans or in landfills and harms the green environment. But, polysaccharide derived 3D bio-plastics offered predictable hardiness e.g., Wyss Institute, USA, has derived chitin based bio-degradable plastic "*Shrilk*" silk protein substitute. Shrilk's hardness and peculiarity offered myriad utilities namely implantable medical devices, bone-tissue gallows, laminated silk fibroin, bio-composts/fertilizer (release nitrogen nutrients), implantable foams, films, surgical closure scaffolds, wound healing, and FDA-approved devices.

2.7.4. *Fabricated chitosan matrix via nano-biotechnology*

Biotechnology can interfaces between science, engineering and technology. Amid, nanobiotechnology is complementary yet untapped science field that can exploits/improves biotechnology and aids to fabricate natural/bio-mimetic nano-structures. Nano-science technology covers and merges biological R&D with various fields as it alters material's parameters including biological, physicochemical and cellular electrochemical responses along with molecular motions. Advanced nano-biotechnology fantastically utilizes natural polysaccharides like chitin/chitosan in such perspectives [1, 2]. Thus, varied matrixes are formulated through chitin/chitosan skeleton owing clinical, biomedical and industrial applicable bio-polymeric prominence like quantum/carbon dots, nano-particle/composite and biomarkers usages in cancer detection. Nano-biotechnology exploited unique and admirable features of chitosan namely biodegradability, biocompatibility, low/no toxicity, antimicrobial activity and low immunogenicity in drug delivery, siRNA/DNA delivery, tissue engineering, and wound healing, biosensors besides theranostics utilities **Figure 1**. Chitosan based many novel nano-materials/devices owns vast beneficial applications to mankind. Strategic chitosan nano-matrixes carriers vital impact on global pharmaceutical use to control drug release due to enhance drugs solubility, superior protein bioavailability, and better uptake of hydrophilic substances across epithelial layers besides great intracellular drug delivery [2–5].

2.7.4.1. Chitosan-carbon dots

Quantum dot are 'nanometer scale' i.e., zero-dimensional particles own semiconducting, optical and electronic characteristics emitted at specific frequency of light that can be adjusted via matter utilized and size, shape and arrangements of dot/particle. Nano-chitonous based carbon quantum dot empowered fluorescent benefits for bio-sensing or imaging due to prominent features namely extremely tune-ability, brilliant water solubility, biocompatibility and better photo-stability [1–4]. Certain chitosan-carbon dots [2–3] yield as smooth, soft films which are robust to UV-visible blockage exploited in biomedical usages imparting low cytotoxicity and excellent biocompatibility with enhanced swelling, better thermal/mechanical properties over pure chitosan film. Amino functionalized fluorescent carbon nitride dots own improved water

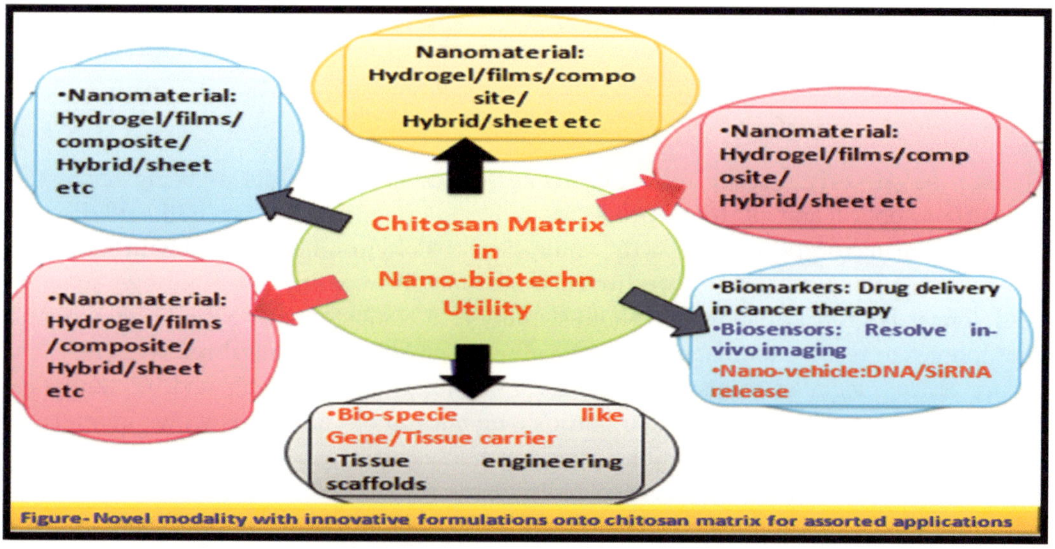

Figure 1. Chitosan matrix in nano-biotechnology.

solubility and strong fluorescent effect as advantageous in medical diagnosis and cancer treatments. Multi-color chitosan-carbon dots exhibited bio-labelling potential with varied bacterial model methods for biomedical usages [2–4]. Solid nano-solar cells are designed via chitin, chitosan and glucose based carbon quantum dot, hybrids as competence of mixed utmost with layer-by-layer sensitizer nanozinc-oxide coating. Such fluorescent nano-crystals/quantum dots can acts as an imaging agent for diseases detection with significant utility. Thus, novel chitosan-zinc sulphide based quantum dots obtained with pH dependent/tuneable optical/electrical properties as probed in pharmaceutical usages [2]. Luminescent chitosan-L-cysteine impregnated cadmium-tellurium films/dots showed antibacterial profile for broad range of biomedical utility. Chitosan-cadmium-telluride quantum dots generated onto indium-tin-oxide coated glass acts as electrochemical biosensor for culprit DNA in chronic myelogenous leukemia treatment or cancer detection. CdS- and/or nano-gold doped chitosan quantum dots are used to formulate antibody immobilization that own brilliant control and bioactive profile compared to other immune-sensors for protein detection studies [2–4]. Such nano-chitosan formations are frequently developed for delivery of safe, effectual plasmid DNA, siRNA, and oligo-nucleotides as gets rapid noticed by genetic materials so as to treat silencing unwanted gene's expression, defects and substituted missing in diseases curing therapeutics [1–5].

Nano-particles can be easily deposited on porous chitosan surfaces to offer homogeneous QDs for multiple active sites adsorption and desorption of hydrogen gas. Certain nano-metallic particles gets integrated into flexible chitosan to exploit for many purpose like catalytic activity, manipulated electronic and chemical characteristics in modern S&T [5]. Polyaniline-chitosan-platinum films are designed on pencil graphite electrode showed brilliant synergic performance in water electrolysis for electro-catalytic hydrogen generation. Thus, chitosan based cheap well-designed gels, films and composites yields via immobilizing desired quantum dots for facilitated hydrogen evolution reaction (HER) and practically skillful hydrogen production with long-term durable electro-catalyst. Flexible organic matrixes of chitosan can be altered for its superior fluorescent, semiconductor behaviour by hyper-branched ligand stabilizer own gelated quantum dots. These fabricated QDs offers solution to multi-responsive novel applications in nano-science for imaging, bio-sensing and drug delivery actions.

2.7.4.2. Chitosan based surface active microcapsules

Chitosan based surface active microcapsule resembles quantum dots as being nano-scale size pertains optical and electronic properties rather intermediate between bulk and discrete. Surface active microcapsules compactly detain electrons/electron-holes varied with size/shape and nature of raw/feedstock to own tuneable features like specific opticity, huge quantum yield, longer fluorescence, enhance photo-stability as advantageous over traditional organic fluorophores in recognition, tagging and imaging in biology and clinical sciences [1–5]. The surface active integrated chitosan skeletons are non-toxic, biodegradable and biocompatible owing liquid-core microcapsule, micro-particle, and macromolecule matrixes own myriad utilities in advanced nanotechnology [6]. Laminar cationic chitosan skeleton owns $\mathbf{-}NH_2\mathbf{-}OH$ linkages that found to assist microcapsule formations via anionic interlocks to implant better biocompatibility and stability in resultant hybrids [6, 7]. Sodium alginate and/or nano-ZnS can micro-capsule chitosan to yield nano-hybrid/nano-gel responsible for many pharmaceutical applications including bio-imaging, bio-labelling and gene/drug release purpose. Flexible chitosan skeleton can catatonically interacts with anionic surface-active surfactants via apparent loss of positive charge to yield insoluble composites or complexes. Such, rational design chitosan based homogenized microcapsule acquired via co-acervation, emulsification, solvent-evaporation, gas-liquid micro-fluidic and layer-by-layer assembly techniques impart improved bioavailability, reproducibility and repeatable drug release/delivery in today's advanced nano-biotechnology [7]. Host-guest interactive and responsive external stimuli sensibility imparts hydrophobic tails/cavities in resultant microcapsules that crucially control its inherent functions as induced via intrusion or doping with surfactants like cyclodextrins and sodium dodecyl sulphate. Strategically controlled/uniform size nano-cadmium sulphide entrapped chitosan quantum dots or microcapsules with stimuli-accountable α-hydrophobic cavities are developed for detection of toxic/hazardous chemicals as beneficial in remediation of ever increasing environmental pollution. Requisite chemical's stimuli-response gets formulated as in rationally controlled chitosan-CdS and/or ZnS quantum dots as liquid-core microcapsules encapsulated with poly(DL-lactide-co-glycolide) showed good fluorescent stability in aqueous condition. Rather usage of surfactants like α-cyclodextrin induces changes/slight influence on shape and fluorescent color in rationally designed chitosan-bends with CdS/ZnS in resulted mono-disperse micro-capsulation. These surfactant induced chitosan stimuli-responsive micro-capsules are cost effective micro-detectors for assorted chemicals than traditional counterparts.

2.7.4.3. *Chitosan based nanoparticle*

Chitosan based nano-particle performs efficient exogenous gene release into primary chondrocytes and imparts immense potential for requisite delivery of therapeutic drugs used in treatment of various diseases. Chitosan skeletal alteration found to enhance transfect ion efficiency by virtue of enhanced self-branching achieved in resultant composites/matrixes/hybrids. Tri-saccharide substitution linear chitosan counterparts compel better gene transfer with intact biocompatibility for cellular uptake with superior stability [3–7]. *N,N,N*-trimethylated chitosan scaffolds encourages *in-vivo* intracellular si-RNA delivery with improved extracellular competence and fine silencing profile along with effective DNA-drug release [1, 8]. Glycol-chitosan scaffolds that can entrap chemo-theraptic like doxorubicin and DOX drugs attenuated utmost si-RNA delivery via surmounts resistance observed in adorn dose-dependent *in-vivo* analysis [8, 9]. Chitosan-poly-D,L-lactide-co-glycolide matrix acts as precise non-viral devices for lot of uses like pulmonary si-RNA release, *in-vitro* H1299 gene silencing and fluorescent protein cell expression [9].

2.7.4.4. *Biosensors & bio-markers*

Sensor receives and responds to signals by converting into magnetic/electrical fields that further detected by an electronic device. Biosensor for biological entities comprises bio-polymeric coalesce and utilizes for physicochemical detection of assorted sensitive natural species like; tissues, cells microorganisms, organelles, enzymes, antibodies and nucleic acids [9]. Biosensor interacts with these analytes and performs recognition/diagnosis based on corresponding interactive evaluation as employed for DNA/RNA, enzymes, antibodies and signal transduced/immobilized tissues [9]. Bio-engineered analytical gizmo aids to offer myriad chitosan based bio-sensores which are beneficial due to uniqueness namely cheap, bio-compatible, eco-friendly, adaptable, portable, high sensitive, intrinsic selective and benign to use in moderately complex environments by virtue of quick responses compared to traditional sensors [10]. Flexible chitosan skeleton recognizes the sample species via immobilization in fabricated matrixes to be exploited for complicated enzyme sensing [11]. Tyrosinase-Fe_3O_4 dope chitosan used to bio-sense/detects certain organic pollutants like catechol onto specifically designed porous nano-iron oxide proactive surfaces [12]. Chitosan diffusion into graphene skeleton induces huge surface area and electrical conductivity in resultant matrix which is used for effectual bio-sensing or immobilization of enzymes and glucose estimation with splendid sensitivity besides durable stability [13]. Nanocarbon dope chitosan yields amperometric matrix to be used for assorted purpose like biosensors, biomarkers, to encapsulate lactase and bio-fuel cells besides as bio-electrochemical devices [14]. Polyaniline-nano-chitosan entrapped creatinine amidino-hydrolase shown good immobilization of CAH enzyme with better stability and durability.

Biomarker is biological indicator use to perform characteristic objective measurements, detections and/or indications for validity of certain phenomenon namely biological state, living organism existence, pathogenic processes and therapeutic intervening pharmacologic responses besides manage cancers and other diseases [15]. Biomarker established doubt in advanced pharmaceutical for facile screening and risk assessment before its diagnosis besides detect diseases with staging, grading and preliminary treatment options for monitoring supplementary therapy [16]. Gold coated chitosan-xanthan biosensor is used for bio-imaging in numerous diseases diagnosis [17] and signal improvement for melanoma. Chitosan-graphene nano sphere marked horseradish peroxidase in α-fetoprotein induce cancer diagnosis with enhanced signal augmentation than electrochemical immuno-sensors [18]. Nano-chitosan biomarkers are preferred detect alpha-fetoprotein and carcino-embryonic antigen with more precision and accuracy over ELISA test. Thus, chitosan integrated biomarkers own stupendous including haemostatic, fungi-static, bacteriostatic, spermicidal, anti-cholestermic and

Figure 2. Chitosan base matrixes for pharmaceutical/clinical and expedient drug delivery.

anti-carcinogenic features fascinated for skillful healings, disease prevention/diagnosis, drug delivery and tissue engineering. Primary —NH_2 link of chitosan smartly offers surface active cationic charge variations (below pH 5.5) achieved via H-bonding imparts adhesion for tissues and subsequent fabricated matrixes offers novel clinically applied products. Some pharmaceutical/clinical expedient utilities of chitosan matrix includes dentistry, orthopedic, ophthalmology, surgical measures, optical and wave guiding utilities as shown in **Figure 2**.

2.7.4.5. Chitosan-doped $Ca(OH)_2$ microcapsules

Liberal —NH_2 groups of chitosan gets protonated in acidic conditions, thus chitosan based microcapsules, micelles and hydrogels can be synthesized to favor varied biomedical/clinical applications. Certain pH-triggered calcium hydroxide coated chitosan hybrids are used for endodontic treatment with long-lasting antibacterial activity against *Enterococcus faecalis* refractory strains. Ethylcellulose-calcium hydroxide coated onto chitosan yields [CS-EC@ $Ca(OH)_2$] microcapsule found to have unique features namely non-toxic, biocompatible, possess superior host immunity, tender controlled drug/gene release besides usage for apical periodontitis in confronted customary root canal therapy. Such CS-EC@$Ca(OH)_2$ microcapsule recuperates innate calcium hydroxide properties which approved osteogenesis effect along with lessening inflammation as beneficial for apical periodontitis in bone defects healing.

Vitapex, calcium hydroxide coated chitosan paste/gels is clinically used to accomplish prominent apical periodontitis effect via stagnant release $Ca(OH)_2$ than other aqueous counterparts. Thus, amid gel vehicles chitosan based microcapsules like Ca (OH)2 loaded propylene glycol-chitosan/agar formulation releases calcium hydroxide via altering gel bases in chronic inflammatory lesion around apex of tooth treatment, besides offered controlled drug delivery with decrease cytotoxicity. Chitosan EC@$Ca(OH)_2$ scaffolds/pH-triggered microcapsule release $Ca(OH)_2$ rapidly in acidic conditions own many significances like inhibits bacterial infections, endorses AP repairs/periapical periodontitis, wide antibacterial activity against *Enterococcus faecalis* refractory strains, great anti-inflammatory profiles and evade bone resorptions. Such, facile and sustainable pH-triggered calcium EC@calcium hydroxide-CS-matrixes own significantly extended antibacterial profile and showed amazingly diminish inflammation to persuade osteogenesis capable in endodontic diseases therapy.

3. Drawbacks and remedies

Amid, all above meritorious usages of chitosan chemistry, it also own certain drawbacks namely as weak basicity at low physiological pH (pka = 6.2) besides, only soluble in aqueous organic acid solution due to protonated —$NH_3^+X^-$ salt formation. Chitosan shows elevated swelling in aqueous medium and carry out unavoidable quick/instant drug delivery, needs to modify/alter chitosan inherent framework. Certain above mentioned demerits in chitosan utility can be trounced by —NH_2 and/or —OH glucosamine derivative formation which solve few complications to cater wide utility. The imperative and characteristics applications of chitosan based bio-polymeric adsorbents established in the environmental pollution remediation is [19] summarized in **Table 1**.

SN	Material types	Practical characteristics		Applications
		Useful property	Adverse property	
1	Nano-filtrations (RO/FO)	Reliable, automated, charge based repulsion, high selectivity, low pressure, costly	High energy, costly, membrane chock, intense polarized, nano-pore dimension	Lessen hardness, color, odor, remove heavy metals, sea water desalination and wastewater treatments
2	Nano-composite membranes	Viable hydrophilic/hydrophobic, huge porosity, more water flux/permeability, thermal stable, mechanically robust, foul resistant, stay at high pressure	Nano-particle leakage, bulk nano-materials needed for oxidation, and composite dependent	Reverse osmosis; eliminate pollutants, composites in ultra-filtration, cartridges nano-fiber composites
3	Self-assemble membrane	Uniform, nano-porous, tuneable size/shape, facile design	Applied on lab/small scales	Ultra-filtrations and process scale up
4	Dendrimer/dendrons (arborol cascade)	Inner hydrophobicity, outer hydrophilicity, water soluble, bio-mimics, easy encapsulation, handy, non-toxic and reusable	Dendrimer/dendron formations are complex multistage processes	Organic/heavy metal removals, biodegradable, biocompatible dendrite
5	Metals magnetic nano-particles	Biocompatible, tuneable colloidal nano-particle, super-paramagnetic, highly recyclable, facile residual separation and huge surface: volume ratio	Stabilization (surface modification needed to enhance its potential	Remediation of hazardous/toxic pollutants, cation/magnetic sensors, nano-beads used for sorption

Table 1. Characteristics of chitosan base membranes in water treatments/purification.

4. Futuristic prospective applications of chitosan matrixes in water purifications

Certain rationally engineered chitosan based adsorbents are vulnerable and flexibly adjustable for facile elimination of contaminants from water/wastewater [14, 20] than compatibly integrated traditional treatments. Bio-materials are beneficially benign to integrate into diversified multifunctional membranes to facilitate both particle retention and contaminant mitigations than conventional materials based water treatments [20]. Further, such chitosan based materials usage own superior process efficiency and high adsorption profile besides adaptable mass scale utility in point-of-use devices [12, 14, 15].

Membrane-based water purifications or treatment processes can address the global challenges of water scarcity and aquatic environmental pollutions [14, 18]. Conservative water purification membranes owe constrained due to inherent limits of conventional materials used in their fabrications [18]. Advanced nano-technology has developed sophisticated methods to control the structural and chemical functionality in chitosan based film, sheet, hydrogel, microcapsule and dots to new classes of materials for water purification [12]. Technologically manufactured chitosan based materials caters the need for futuristic advanced water purification/treatments. Such molecularly designed well selective and focus materials are highlighted by surface modification to minimize interfacial interactions, inherent limitations and enhance efficiency over customarily used materials as in **Table 2**.

Utility	Nature of work		Features
Energy	Sunlight conversion (like DSSCs, PBs)		• Efficient light-harvesting, especially in biomaterials replica or biocomposites examples; • fast charge separation, high current density; • high gas permeability; • high storage density; • fast electron and ion transport; • small resistance
Life sciences	Engineered/designed Biomaterials/films/composites		• Biocompatibility; • promoting cell adhesion; • good mechanical property; • controlled shapes/sizes
Chemical sciences	Pre-concentration device For separation and adsorption	Bioreactors	• High permeability; • homogeneous flow-through pore structure; • control pore structures and surface properties; • used monolithic column

Table 2. Applications of chitosan matrixes.

5. Conclusion

Chitosan biopolymer is preferred for biomedical and pharmacological promising purposes as its safe/harmless besides impart successful drug delivery/releases. Muco-adhesive chitosan improved abode span and consequently provide drug's bioavailability in target specific carriers. Certain fillers like ZnO, ZnS and TiO_2 chosen to intrude into chitosan skeleton to yield resultant matrixes/hybrids/composites with widen utilities in clinical/pharmaceuticals. Systematically performed R&D in chitosan based scaffolds/matrixes owing interactive filler/dopant added to improve its applicability which opens subsistence revolutionary and advanced medical usages.

Nanobiotechnology integrated technology and complementary strengths of bio-molecule chitosan with the nano-electronics to own assorted applications/outcomes as biosensor and biomarker proliferation as high-resolution devices (for *in-vivo* imaging) and drug delivery treatment in cancer respectively. Further advance biotechnology exploited excellent biological characters of chitosan so as to introduce novel modality with innovative formulations namely nanovehicles for gene/DNA/SiRNA delivery and quantum dots/nanobio-composites for various diseases diagnosis, generating new therapeutic techniques besides development of tissue engineering scaffolds. Thus, nano-biotechnology skillfully explored chitosan matrix to design and device never-ending clinical and scientific applications for the benefit of human besides environment/nature.

Acknowledgements

The author is thankful to the Head, Department of Chemistry, R.T.M. Nagpur University, Nagpur, for laboratory facilities and to the Vice Chancellor, Nagpur University, Nagpur, for the sanction of a research project under University Research Project Scheme, No. Dev/RTMNURP/AH/1672 (9) dated 24 September 2016.

Author details

Rajendra Sukhadeorao Dongre

Address all correspondence to: rsdongre@hotmail.com

Department of Chemistry, RTM Nagpur University, Nagpur, Maharashtra, India

References

[1] Dongre RS. Biological Activities & Application of Marine Polysaccharides. Vol. 1. Croatia: In-Tech Open; 2017. pp. 181-206. DOI: 10.5772/65786

[2] Jain1 T, Kumar S, Dutta PK. Chitosan in the light of nano-biotechnology: A mini review. Journal of Biomedical Technology and Research. 2015;**1**(1):101-107

[3] Bănică F-G. Chemical Sensors & Biosensors: Fundamentals & Applications. Vol. 176. UK: John Wiley Sons; 2012. ISBN: 978-0-470-71067-8

[4] Jianc H, Su W, Caracci S, Bunninc TJ, et al. Optical wave guiding and morphology of chitosan thin films. Journal of Applied Polymer Science. 1996;**61**:1163-1171

[5] Burkatovskaya M, Tegos GP, Swietlik E, Demidova TN, P Castano A. Use of chitosan bandage to prevent fatal infections developing from highly contaminated wounds in mice. Biomaterials. 2006;**27**:4157-4164

[6] Kurita K. Chitin & chitosan: Functional biopolymers from marine crustaceans. Marine Biotechnology. 2006;**8**(3):203-226

[7] Mourya VK, Inamdar NN. Chitosan-modifications and applications: Opportunities galore. Reactive and Functional Polymers. 2008;**68**(6):1013-1051

[8] Sashiwa H, Shigemasa Y. Chemical modification of chitin & chitosan 2: Preparation & water soluble property of N-acylated or N-alkylated partially deacetylated chitins. Carbohydrate Polymers. 1999;**39**(2):127-138

[9] Dutta PK, Dutta J, Tripathi VS. Chitin and chitosan: Chemistry, properties & applications. Journal of Scientific and Industrial Research. 2004;**63**:20-31

[10] Peter MG, Domard A, Muzzarelli RAA. Advances in Chitin Science. Vol. IV. Potsdam, Germany: Universität Potsdam; 2000. p. 395. ISBN: 3-9806494-5-8

[11] Hudson SM, Jenkins DW. Chitin & Chitosan. Encyclopedia of Polymer Science and Technology-1. 3rd ed. New York: Wiley Interscience. DOI: 10.1002/0471440264.pst052

[12] Dutta PK, Ravikumar MNV, Dutta J. Chitin and chitosan for versatile applications. Journal of Macromolecular Science-C, Polymer Review. 2002;**42**(3):307-354

[13] Noipa T, Ngamdee K, Tuntulani T, Ngeontae W. Cysteamine CdS quantum dots decorated with Fe^{3+} as a fluorescence sensor for the detection of PPi. Spectrochimica Acta—Part A: Molecular and Biomolecular Spectroscopy. 2014;**118**:17-23

[14] Yan J-J, Wang H, Zhou Q-H, You Y-Z. Reversible and multisensitive quantum dot gels. Macromolecules. 2011;**44**(11):4306-4312

[15] Hardison D, Deepthike HU, Pathirathne T, Wells MJ. Temperature-sensitive microcapsules with variable optical signatures based on incorporation of quantum dots into highly biocompatible hydrogel. Materials Chemistry. 2008;**18**(44):5368-5375

[16] Sá-Lima H, Caridade SG, Mano JF, Reis RL. Stimuli-responsive chitosan-starch injectable hydrogels combined with encapsulated adipose-derived stromal cells for articular cartilage regeneration. Soft Matter. 2010;**6**(20):5184-5195

[17] Thongngam M, McClements DJ. Influence of pH, ionic strength & temperature on self-association & interactions of sodium dodecyl sulfate in the absence & presence of chitosan. Langmuir. 2005;**21**(1):79-86

[18] Chen Y, Yao R, Wang Y, Chen M, Qiu T, Zhang C. CdS QDs-chitosan microcapsules with stimuli-responsive property generated by gas–liquid microfluidic technique. Colloids and Surfaces. B, Biointerfaces. 2015;**125**:21-27

[19] Li X, Yang Z. Chitosan-decorated calcium hydroxide microcapsules with pH-trigger release for endodontic application. Journal of Materials Chemistry B. 2015;**3**:8884-8891

[20] Kayan DB, Koçak D, İlhan M. The activity of PAni-Chitosan composite film decorated with Pt nanoparticles for electrocatalytic hydrogen generation. International Journal of Hydrogen Energy. 2016;**41**(25):10522-10529

Chapter 3

Extractive Fermentation Employing Ion-Exchange Resin to Enhance Cell Growth and Production of Metabolites Subject to Product or By-Product Inhibition

Fadzlie Wong Faizal Wong, Murni Halim and Arbakariya B. Ariff

Additional information is available at the end of the chapter

http://dx.doi.org/10.5772/intechopen.76879

Abstract

In recent years, commercial production of proteins and metabolites from microbial fermentation for industrial applications has increased significantly. Innovative approaches are directed towards the improvement of the conventional batch fermentation method and the segregated downstream processing of target product to improve the overall process efficiency and to ensure that the process is economically viable. Feedback inhibition is a common problem faced during fermentation process when the concentration of end-product/by-product reaches a certain level. The excessive accumulation of end-product/by-product in the culture may inhibit the growth of cell and represses the secretion of target metabolite. In the production of many fermentative products such as antibiotics, amino acids, and fungal metabolites, a serious problem of feedback inhibition is often encountered. Cultivation of lactic acid bacteria and recombinant bacteria is usually subjected to by-product inhibition. Hence, extractive fermentation via *in situ* ion-exchange-based adsorptive technique is a possible approach to be used industrially to mitigate feedback inhibition, aimed at enhancing fermentation performance. In this chapter, advances in this area were presented. Strategies to overcome problem related to product/by-product inhibitions by this technique via dispersed, external, and internal resin system, and the general methodology in the implementation of the technique were also discussed.

Keywords: extractive fermentation, ion-exchange resin, product/by-product inhibition, feedback inhibition, *in situ* extraction

1. Introduction

Fermentation generally is a metabolic process, in which a microorganism (e.g. yeast or bacteria) converts a substrate (e.g. starch or sugar) to microbial metabolic products, whether in a shake flask or bioreactor system. These metabolic products can be divided into: (1) primary metabolites, that is, compounds that are produced during the logarithmic phase of microbial growth, and they are essential for growth (e.g. alcohol, lactic acid, and amino acids) and (2) secondary metabolites, that is, compounds that are produced during the stationary phase of microbial growth, and they generally do not have any function in growth except maybe for survival function (e.g. antibiotics, anticancer agents, pigments, and organic acids).

Under normal circumstances, in the context of industrial biotechnology, there is usually only one target product in fermentation; therefore, this means that the rest of the simultaneously produced metabolites (as highlighted above) are instead considered as by-products. For example, during the fermentation of recombinant *Escherichia coli* to produce a targeted product periplasmic interferon alpha-2b, the acetate ion (which is also commercially valuable) produced is considered to be the by-product of the fermentation [1]. The primary concern on the accumulation of products/by-products in culture medium is the possibility of inhibiting cell growth and subsequently suppressing the production of targeted product.

The examples of products/by-products that exert feedback inhibition and their general mechanism of inhibition are:

i. Alcohols

Feedback inhibition is best exemplified by acetone-butanol-ethanol (ABE) fermentation, where butanol concentration as low as 1% w/w severely inhibit cell growth, resulting in low titers [2]. Alcohol like ethanol does not cause protons to leak through membrane; however, it does cause structural and function alterations of cellular membrane. The rate of substrate utilisation in alcohol-treated cells is lower relative to control. This directly points to the theory that the inhibition is a result of irreversible denaturation of membrane-associated glycolytic enzymes [3].

ii. Organic acids

In most of the *in situ* organic acid removal studies reported so far, the organic acids can be either the desired product or by-product of fermentation for producing recombinant protein or enzyme. The accumulation of organic acids in cultivation medium generally has an adverse effect on productivity [3]. Organic acids act as uncouplers that allow proton (H^+) to enter cell from medium and counteract the action of proton pump [4]. This results in the interference in the difference between internal and external pH (ΔpH). For every molecule of acid entering the cell, one H^+ is internalised, as the species permeating to the inside is neutral acid (HA) and the one leaving the cell is anion (HA_{2}-). The attempt of cell to restore this imbalance by expending more energy (in the form of ATP) at membrane level rather than for biosynthesis thus could explain the mechanism behind the inhibition by organic acids.

iii. Antibiotics

Antibiotic is a type of secondary metabolite that can also be lethal towards its own producer strain [4]. In comparison with other types of inhibitors mentioned above, which are multi-target inhibitors, antibiotic usually interacts specifically towards cell, owing to the unique inhibitor-receptor structural complementation.

iv. Others

Besides the above-mentioned inhibitors, the accumulation of some metabolites such as pigment [5], anticancer agent [6] and so on, up to a certain concentration threshold, has also been reportedly suppressed their own production. Nevertheless, their mechanisms of inhibition are still not very clear.

In view of this, in order to improve fermentation performance (yield, productivity, and final concentration of biomass or secondary metabolites), beside the requirement to optimise fermentation conditions (e.g. inoculum density, pH, temperature, aeration and mixing, and concentration of nutrients) and to alleviate substrate inhibition effect (if any)—where process control via fed-batch operation and change of substrate type can be implemented, one must also address the problem of product/by-products inhibition.

So far, various strategies have been used to address the problem. Genetic engineering is one of the approaches used to overcome the feedback inhibition. The primary aim of this technique is to develop a product/by-product tolerant strain, which may be achieved, specifically, via the modification of the product receptor on the cell membrane, overexpression of protein, deletion of metabolic pathway, and so on [4, 7]. Besides, extractive fermentation (interchangeably known as *in situ* removal/recovery), which basically is an integration of fermentation with product/by-product removal, has been extensively studied in order to mitigate the problem of feedback inhibition, to improve the productivity of bioprocesses by converting the conventional multi-step methods to one-pot process and to address product decomposition after its formation [8]. **Table 1** lists some of the techniques (which have been categorised based on the basis of product extraction) that have been employed previously in extractive fermentation. The selection of the techniques is basically depending on the nature of the target product/by-product. In addition, Hecke et al. designed a heuristics for selecting the appropriate technique based on the properties of target compound (inhibitory, stability, volatility, and charge) [7] (**Figure 1**). For volatile products such as the fermentatively produced acetone-butanol-ethanol, techniques such as gas stripping, distillation, and membrane-based technology, pervaporation have been proposed [2]. For organic acids, separation based on adsorption, electrodialysis, and solvent extraction has been widely reported [3]. Moreover, the overall developments in these extractive fermentation techniques have also been extensively reviewed earlier [2, 7, 9, 10].

On a separate note, ion-exchanger is an inert support medium that is covalently coupled to positive/negatively charged functional groups, of which oppositely charged ions are bounded. The counter ions will be exchanged with like-charge ions in a sample (provided that their charge magnitude is larger than the one possessed by the bounded-ions). Extractive fermentation can be performed either by adding an ion-exchange resin into a fermentation broth or by passing the broth through an adsorbent-filled column, where inhibiting product/by-product

No.	Basis of *in situ* extraction	Specific techniques	Examples of product
1	Target product immobilisation	Adsorption onto polymeric matrices such as ion-exchangers, activated carbon, zeolites, and cellulose	Ethanol, salicyclic acid, cycloheximide, anthraquinone, alkaloids, monoterpenes, tissue plasminogen activator
2	Extraction into a second liquid phase	Solvent extraction, aqueous two-phase system, and supercritical fluid	Alcohol, organic acids
3	Size selective permeation	Membrane-based techniques	Microbial, mammalian, and hybridoma cells
4	Evaporation	Vacuum fermentation, gas stripping, and evaporation	Ethanol, butanol
5	Change in solubility of target product	**i.** Formation of complex via the addition of soluble reagent	Acetaldehyde
		ii. Crystallisation of target product Schiff's base formation	Carboxylic acids, antibiotics, amino acids, steroids

Table 1. Categorisation of *in situ* extraction techniques made by freeman et al. but with slight modification [11].

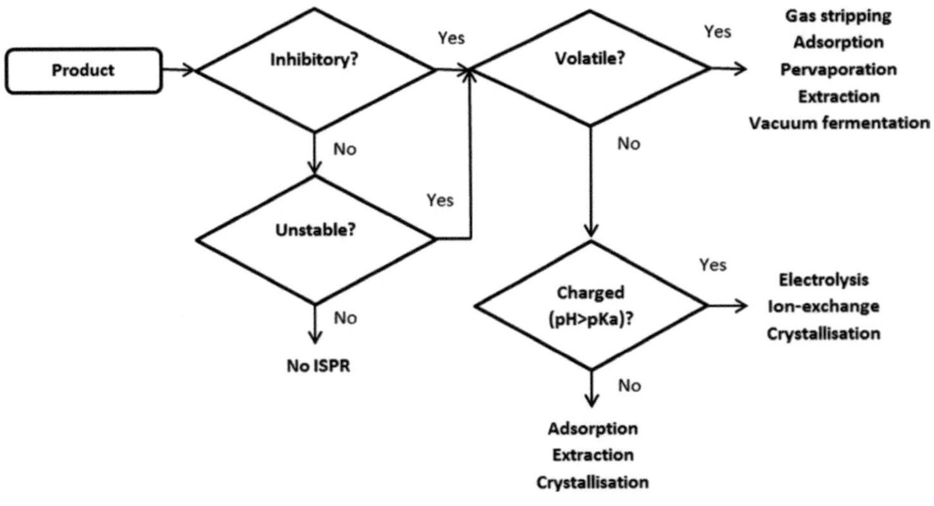

Figure 1. Heuristics for selecting suitable *in situ* product removal technique based on the nature of target product [7].

will be captured by the resin, thereby maintaining the optimal culture condition for cell growth or production of metabolites (**Figure 2**). The advantages of using ion-exchange-based separation have been extensively elaborated previously [6]. The sequestration of product/by-product helps to mitigate the feedback inhibition, improving fermentation performance. Besides, for fermentation products that are prone to degradation (e.g. fragrances and flavours), their stability in the culture medium can also be improved. In comparison with other organic solvent-based extraction, ion-exchange-based techniques may also increase the stability of product by reducing the length and scale of exposure of such products to organic solvent.

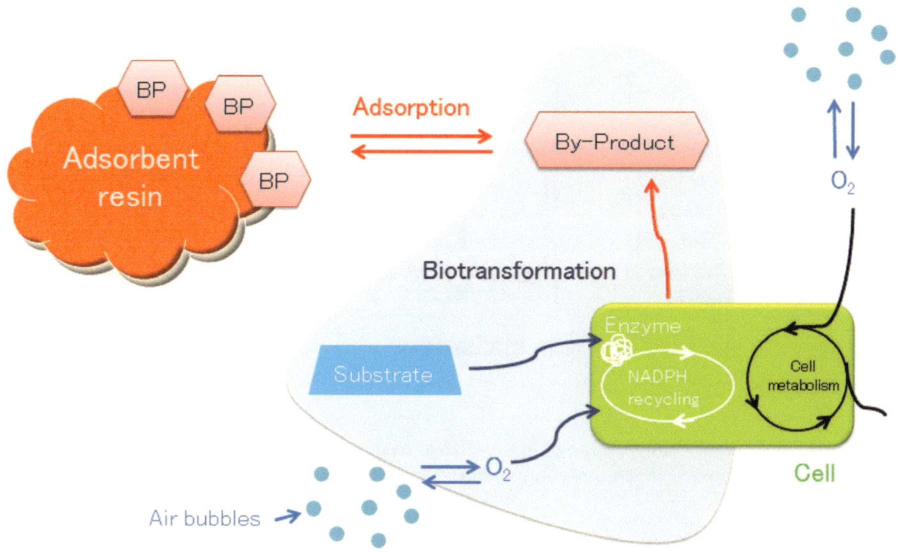

Figure 2. Schematic diagram representing the concept of extractive fermentation via *in situ* removal of by-products (BP) to enhance cell growth and production of metabolites.

The use of resin also offers an economic benefit as it is reusable. Additionally, its usage is also environmentally friendly since only a minimum consumption of organic solvent is required during preparation and elution stages. Moreover, this technique is also relatively simple to operate as resin separation following a fermentation process is relatively easy and therefore will not require the use of any high-end instrument.

In this chapter, the reports on the application of ion-exchange resin in addressing the problem of feedback inhibition from product/by-product in microbial fermentation, in different configurations: dispersed, internal, and external system are reviewed. In addition, the general methodology in developing an extractive fermentation based on the use of an ion-exchange resin is discussed. Finally, a perspective on the application of this technique is also presented.

2. Applications of ion-exchange adsorption resins in extractive fermentation

The adsorption of target product/by-product by ion-exchange resin in an extractive fermentation can be carried out either within a bioreactor (internal system) or by circulating a fermentation broth through an external column that is packed with the adsorbent (external system) (**Figure 3**). Additionally, in the former system, the resins can either be trapped in a compartment housed inside the bioreactor or dispersed freely in the culture.

2.1. Dispersed resins

The feasibility of improving the growth of attenuated gdhA derivative *Pasteurella multocida* B:2 (which is to be used as bacterial vaccine for animal) by removing the by-product

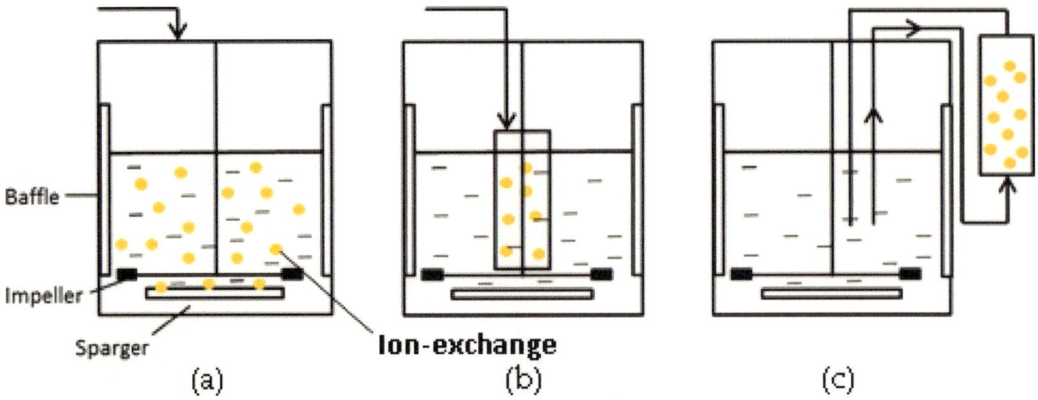

Figure 3. Schematic diagram of configuration of ion-exchange resin systems in extractive fermentation: (a) dispersed resins in culture, (b) trapped resins in internal column and (c) external column.

ammonium (NH_{4^+}) using cation-exchange resins was previously investigated [12]. The accumulation of NH_{4^+} inside cells generally results in intracellular pH change, and subsequently interrupts the activity of cytosolic enzymes [13]. Three types of resins: the weak acid cation-exchanger - Amberlite IRC86, the strong acid cation-exchangers - Amberlite IR120 H, and Dowex DRG8 H were examined. During a sorption isotherm experiment, it was determined that the highest maximum binding capacity, q_{max} (0.863 g/g) and the lowest dissociation constant, K_d (0.015 g/L) were obtained with Amberlite IRC86. Although the other two strong cation-exchangers showed high ammonium adsorption compared to IRC86, but lower cell viability was observed in their systems, which might be due to the co-removal of cation nutrients in the culture media. Additionally, when the amount of ammonium was increased from 300 to 1500 g/L, the desorption efficiency of ammonium from the resins (using 2.5 M HCl), reduced by 40 to 50%. In a shake flask culture (100 mL), the cell viability drastically reduced (5.5×10^9 times) when the concentration of IRC86 was increased from 0 to 30 g/L. The optimum resin concentration was determined to be 10 g/L, where maximum NH_4 concentration of 536.40 ± 10.85 mg/L (41% removal) and cell viability of 7.2×10^{10} (equivalent to 13-fold improvement compared to resin-free system) were obtained.

A reduction in cell yield is usually inherent in the accumulation of organic acid such as lactic acid, acetic acid, and succinic acid. Acidification of medium, followed by cytoplasm (as lactic acid diffuses into the cell), results in failure of proton motive forces, which ultimately damaging the cells [14]. However, this acidification problem may not simply be overcome by a pH control strategy, that is, addition of NaOH. It was highlighted earlier that the osmotic pressure of medium escalated due to the additional ions from NaOH, resulting in reduced cell growth [15, 16]. Therefore, Othman et al. combined the strategy of using fed-batch mode operation with an extractive fermentation of probiotic *Pediococcus acidilactici* using anion-exchange resin, Amberlite IRA 67 [16]. About 55.5 times increment in maximum viable cell concentration was achieved in the fed-batch extractive fermentation system compared to a batch mode. The maximum viable cell concentration, yield, and productivity achieved in the ion-exchange resin-added fed-batch fermentation, was 9.1, 8.5, and 8.6 times higher compared to a control system (without resin), respectively. Lactic acid accumulated in the former system was also lower (8.78 g/L) than system without resin (9.62 g/L). The fed-batch operation improved the fermentation performance by: (1) overcoming the substrate inhibition effect by maintaining the glucose concentration in the culture at below the inhibition level and thereby directing the glucose metabolic flux towards cell growth rather than lactic acid production and (2) enhanced the fermentation performance by removing the inhibitory by-product (i.e. lactic acid) following the *in situ* addition of ion-exchange resin.

Furthermore, an accumulation of acetate in fermentation medium also causes retardation of cell growth (due to disruption in transcription-translation machinery, stress response, and regulation in cell), and subsequently, affecting the productivity of desired metabolites or the expression of recombinant proteins (in the case of recombinant strain). *In situ* adsorption of acetate by anion-exchange resins in *E. coli* culture for producing periplasmic human interferon-alpha2b (PrIFN-α2b) was previously studied in shake flask (250 mL) and bioreactor (2 L) systems [1]. Selection of anion-exchange resin was made out of two strong base anion resins (Amberlite IRA-900 and Dowex Marathon MSA) and three weak base anion-exchange resins (Amberlite IRA-96, Diaion WA30, and Dowex M43). The selection was made based on the affinity between the resins and acetate that was determined from a sorption isotherm experiment (the affinity is indicated by the value of dissociation constant, K_d, where the lower the value, the higher the affinity). In comparison, the use of weak anion-exchange resin gave higher cell growth and PrIFN-α2b expression than the strong base anion-exchange resins. Optimisation of ion-exchange resin load for all of the resins tested was also carried out. Interestingly, for all of the resins, the cell growth and PrIFN-α2b expression were found to decrease when their concentrations were increased from 20 to 80 g/L due to higher accumulation of acetate in the batch culture.

The co-adsorption of anion nutrients such as Cl^-, SO_4^{2-} and PO_4^{2-} might be the cause of the suppressed growth and PrIFN-α2b expression. WA30 was selected and used for further experiment, as the highest cell and PrIFN-α2b concentrations were obtained, that is, 8.65 ± 0.13 g/L and 566.5 ± 4.36 μg/L, respectively. Additionally, the lowest concentration of acetic acid accumulation was also recorded (i.e. 3.49 ± 0.26 g/L) in the WA30 culture. Based on optimisation using response surface methodology (RSM), an optimum loading of 12.2 g/L was determined, where the maximum yield of 501.8 μg/L was achieved, in accordance with the predicted yield (i.e. 507 μg/L).

The effect of ion-exchange resin addition on the physiology of the cells was also investigated. Based on scanning electron microscope (SEM) analysis, the morphology of the cell grown in resin-added medium was found to be identical to the one from the control system, which is a long rod shape. However, as visualised via transmission electron microscope (TEM), a higher amount of inclusion bodies aggregates was detected in the cells grown in high resin-load culture system (i.e. 80 g/L) compared to one from low resin-load system (i.e. 20 g/L). This finding further explained the lower cell growth and PrIFN-α2b expression in the high-resin load system. Finally, it was also found that the production of PrIFN-α2b by *E. coli* in stirred tank bioreactor (STB) with *in situ* addition of WA30 resin was enhanced by 1.8-fold (578.8 μg/L) compared to system without resin (318.4 μg/L). A reduction in acetate concentration (28%) was obtained with the addition of resin after 26 h of fermentation (**Figure 4**).

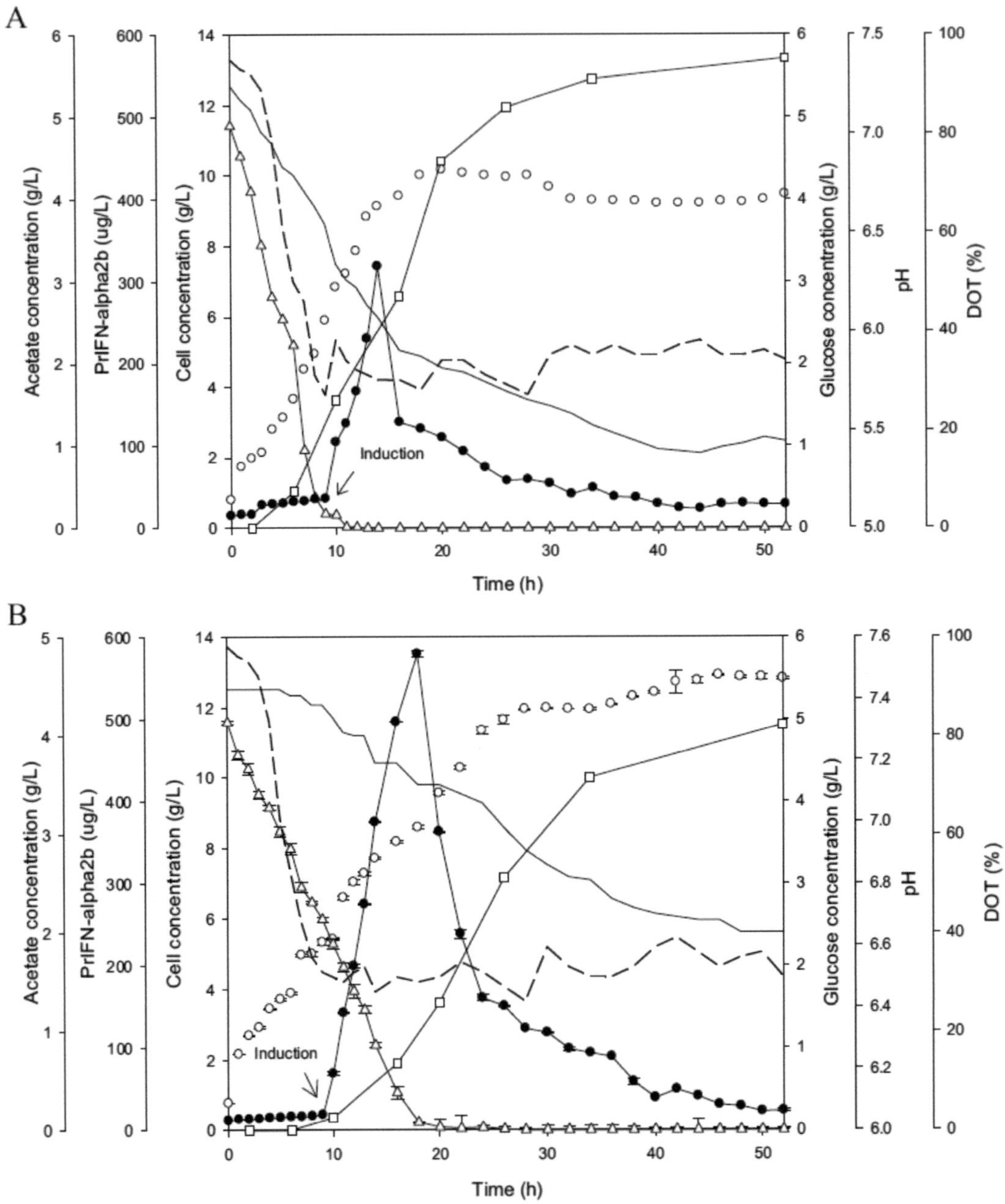

Figure 4. The time course of recombinant *E. coli* fermentation for the production of PrIFN-α2b in 2 L STB: (a) control system (resin-free), (b) with *in situ* addition of resin [1]. Symbols: (•) PrIFN-α2b concentration, (○) cell concentration, (□) acetate concentration, (Δ) glucose concentration, and (--) dissolved oxygen tension.

Another application of ion-exchange resin for removing acetic acid in a culture was reported by Chen et al. [17]. Acetic acid accumulation was also found to have a critical influence on the production of human epidermal growth factor (hEGF) by recombinant *Escherichia coli* JM101.

Growth of *E. coli* was not occurred at 20 g/L of acetic acid concentration. The weak basic IER A-D3-1 macroporous resin was added to a shake flask culture and a batch culture in a 2.5 L bioreactor. In fermentation using bioreactor, where 33 g/L of resin was used, a 10% increment in expression level was obtained, relative to resin-free system. The resin was found to be non-toxic to the cell, albeit a minimum decrease in growth and production of hEGF was observed. This phenomenon was attributed to the co-removal of nutrient components such as CH_3COO^-, PO_4^{3-}, SO_4^{2-}, OH^-, Cl^- and other charged amino acids, as a result of their non-specific complexation with the resin.

In the production of the self-toxic antifungal agent, cercosporamide by fungal culture LV-2841, it was found that no significant improvement in yield was obtained following extensive optimisation on medium (types and concentrations of carbon (C) and nitrogen (N) sources) and fermentation conditions (aeration and temperature). Therefore, an extractive fermentation via *in situ* addition of ion-exchange resin to adsorb the product cercosporamide was done. The study was carried out in various fermentation scales, ranging from 100 μL to 10 L. Over 100-fold increase in production titre was achieved when the potato dextrose broth was supplemented with 1–3% w/v wet Diaion® HP20 resin or AmberliteTM XAD-7.

Moreover, the problem of low titre (less than 10 mg/L) was encountered in the fermentation of *Burkholderia thailandensis* E264 for producing thailandepsin A, a natural anticancer analog [6]. Low titre can deter further preclinical/clinical development of the natural drug. Thus, *in situ* addition of a polyaromatic adsorbent resin, Diaion HP-20 to the culture, coupled to a systematic optimisation of fermentation conditions was carried out in order to increase the product titre. Optimisation of Diaion HP-20 concentration was initially done by single factor experiments. The addition of resin at 4% (w/v) (after 12 h of fermentation) resulted in 91% increment in thailandepsin A production. Nevertheless, reduced production of the thailandepsin A was observed with further increase in resin concentration. It was presumed that at high resin loads, bacteria's nutrients adsorption was affected due to the resin attachment to the bacteria and/or nutrient. Additionally, there might also be larger shear forces as a result of limited space. Furthermore, at higher dissolved oxygen concentrations, which were resulted by the addition of the resin, the cell growth might as well have been inhibited.

Previously, an attempt to increase the production of teicoplanin, an antibiotic produced by *Actinoplanes teicomyceticus* in a 5 L STB using ion-exchange-based extractive fermentation has been reported [18]. In a preliminary experiment, the threshold level for the cell growth and production inhibition was determined to be from 10 to 25 mg/L. In selecting the suitable resin for subsequent experiment, resins/adsorbents: Diaion HP-20 (adsorption by ionic properties), Amberlite XAD-16 (adsorption by hydrophobicity), charcoal (adsorption by polar properties), and silica gel (adsorption by decolourisation of teicoplanin) were added to the fermentation broth at a concentration of 5% (v/v) at the inoculation stage and incubated for 120 h. The desorption step was performed using 80% methanol. Among all of the resins tested, Diaion HP-20 exhibited the highest adsorption and desorption performances followed by XAD-16. Although a high adsorption was achieved by charcoal, no recovery was achieved during desorption, indicating a strong adsorption between the teicoplanin and adsorbent. Since Diaion HP-20 is an ion-exchanger and a polyaromatic resin polymerised with polystyrene and divinylbenzene, the capture of the teicoplanin by the resin was therefore based on ionic and hydrophobic interactions. Under a resin concentration of 5% w/v, which was supplemented to the culture at inoculation stage, a teicoplanin concentration of 134 mg/L (about 4.2-fold increment) was achieved after 120 h of fermentation.

Beside inhibitory problem, enzymatic degradation of product also commonly occurs during fermentation. One example of such product is pristinamycin, an antibiotic produced by *Streptomyces pristinaespiralis* [19]. The feasibility of increasing the productivity of pristinamycin production by *in situ* addition of resins to the culture in Erlenmeyer flask and 3 L STB has been examined. The types of resins that were being investigated include HP-20, XAD-16, JD-1, and HZ-817. Based on the outcome of batch binding experiment, JD-1 was selected to be used for the subsequent study, as high adsorption capacity (25.7 ± 1 mg/g) and desorption ratio (84.2 ± 1%) were achieved. In Erlenmeyer flask cultivation, a reduction in dry cell weight (from 16 to 13 mg/mL) was observed, when the resin load was increased from 0 to 16% w/v. The pristinamycin production was also increased by 1.83-fold when JD-1 was added at the 20 h of fermentation compared to addition at just after inoculation, because the onset of pristinamycin synthesis occurred on the 20 h. When the ion-exchange system was adapted to a 3 L bioreactor, the pristinamycin synthesis reached 0.8 g/L, which was about 1.25-fold increase over a resin-free cultivation.

2.2. Internal adsorption column

An internal adsorption column configuration (in a bioreactor) afforded the advantage of avoiding the influence of shear force from impeller that could have a detrimental effect on the resin. Although the same benefit can be attained from the use of an external column system, additional advantages can also be found with this design over the external variant, where better environmental conditions, such as pH, temperature, dissolved oxygen tension (DOT), aeration rate, and mass transfer rate can be achieved [20]. Next, product recovery can be done within one bioreactor as the fermentation (*in situ* product recovery), and this improves overall productivity, as well as providing a simpler operation. Bae et al. reported on the development of a novel internal column system to recover prodigiosin-like pigment produced from the fermentation broth of *Serratia* sp. KH-95 [20]. The bioreactor was designed with a second compartment made by internal 316 stainless steel filter and perforated support for a HP-20 adsorbent. **Table 2** summarises the performance of the system relative to external column, internal-dispersed, and without resin (control) systems. Notably, this novel system afforded 1.8-fold higher in pigment productivity than that obtained in the external column-bioreactor system.

The phenomenon of cell adsorption on the resin was reportedly encountered in dispersed and external resin systems, as the pigments were still attached to the cell wall. It was also highlighted that the conditions (e.g. pH, DOT, substrate concentration, etc.) in the external column and internal bioreactor medium were possibly varied. Thus, this might also resulted in the lower performance in the external resin system since the cell is recycled throughout the whole system.

2.3. External adsorption column

The main drawback inherent to dispersed resin system is resin attrition caused by the shear force of impeller(s) and the clogging of the valves and shaft seals in bioreactor [20]. This

No	System	Recovered product concentration (g/L)	Maximum cell concentration (g/L)	Pigment productivity (g/L/h)
1	Internal column	13.1	<1.0 (6)	0.44
2	External column	7.2	3.0 (15)	0.24
3	Dispersed adsorbent	9.8	<0.5	0.33
4	Resin-free system (control)	5.2	NA	NA

All adsorbent systems were carried out in 2.5 L bioreactor (1 L working volume), while the volume of the resin used was 100 mL. Acidified methanol was used to desorb the resin. The number in the bracket indicates the time taken to reach the maximum cell concentration (in hour). NA: data not available.

Table 2. Comparative evaluations of different resin application configurations: Internal, external, and dispersed, relative to resin-free system (control) in the fermentation *Serratia* sp. KH-95 to produce prodigiosin-like pigment [9].

effect can be pronounced in an aerobic fermentation, where an agitation is usually required. Thus, an external adsorption system is proposed in order to overcome any possible effect of shear force on cells from the collision of resins. Generally, in the concept of external resin extractive fermentation, a separation unit (resin-loaded column) is connected to a bioreactor externally, and a fermentation broth is passed through the column for capturing of product/by-product, subsequently, the broth is returned to the bioreactor for reuse. Furthermore, the broth reuse also afforded further advantages, that is, higher conversion of nutrient/substrate and the reduction in input water requirement [21].

2.3.1. Integrated bioreactor-packed bed column

In situ citric acid recovery in *A. niger* W1-2 fermentation by anion-exchange resin was previously developed [21]. The resin-loaded packed-bed column (4.0 × 50.0 cm) was connected to a 1 L bioreactor (0.7 L working volume) system. The broth was passed through the column only after 3 days of fermentation, at a frequency of once a day and a flow rate of 5 mL/min. The resin was found to be not attracted towards nutrients (i.e. residual sugar and NH_4Cl) except for KH_2PO_4. In comparison with a conventional fermentation, reduction in fermentation time (6 days), increment in productivity (1.6-fold) and sugar conversion (12.6%) were achieved by the integrated system.

A study to overcome end-product inhibition during the fermentation of *Serratia* sp. KH-95 for producing prodigiosin-like red pigment using ion-exchange resin was previously carried out [5]. Purified pigment was added to a shake-flask culture of *Serratia* to confirm the effect of product inhibition. Selection of resin (among HP-20, SP-850, and XAD-16) was made based on its adsorption performance in fixed bed column mode, where the pigment-containing culture was loaded from the top of the column. Although SP-850 showed the highest adsorption capacity (about 11.8% and 38.1% higher than HP-20 and XAD-16, respectively), it exhibited the lowest desorption ability (about 15% and 20% lower than HP-20 and XAD-16, respectively). Thus, HP-20 was selected and used for subsequent studies. The effect of resin addition time (0–15 h after inoculation) on the fermentation performance was also investigated in shake flask fermentation. Overall, it was found that cell growth was suppressed once the resin was added. Notably, the cell growth only reached 75% of the control (resin-free system) when the resin was added during inoculation. However, the growth steadily increased to 95% at 10 h of resin addition time. This phenomenon was attributed to the reduction in casein content in the culture (about 23.5–47.1% reduction relative to control) due to undesired adsorption to the added resin that resulted in amino acids depletion in the medium. On the other hand, the production of pigment increased correspondingly with the increase of addition time. The maximum concentration of pigment (5.85 g/L) was obtained at 10 h of addition time (40% increment relative to control).

Furthermore, when resin amount was increased from 5 to 20% (v/v), cell growth and casein concentration were found to reduce. Nevertheless, 10% (v/v) was chosen as the optimum load, as the maximum production of pigment was obtained. Finally, the extractive fermentation was adapted to a bioreactor scale (5 L), which was connected to a packed-bed column. Similar to shake flask culture, the highest production was also achieved with 10% v/v resin load. Overall, 31% increment was afforded in the integrated system compared to a resin-free bioreactor system.

2.3.2. *Integrated bioreactor-expanded bed adsorption*

Tan et al. designed an integrated STB-EBA system for removing acetate to enhance the production of PrIFN-α2b by *E. coli* [22]. In comparison with a packed-bed column, EBA system allows for direct handling of biomass-laden fermentation broth without the need for any intermediate clarification method. Due to the bed expansion characteristic of EBA, where a large void volume is created, cell/particulate will pass through the column, while the target compound will be captured by the adsorbent in the column. The system was set-up based on a 2 L STB (1 L working volume), that was coupled to an UpFront FastLine 10 column (UpFront Chromatography A/S, Copenhagen, Denmark) (i.d 10 mm) packed with a WA30 anion-exchange resin (**Figure 5**). The effects of EBA parameters: amount of resin loaded (0–12 g/L) (and hence, the sedimented bed height, H_0 from 0 to 150 mm), column superficial linear velocities (from 240 to 900 cm/h), and culture viscosity (3.2–113.9 mPa.s) on the fermentation performance were first studied. It was found that at 150 mm sedimented bed height (12 g/L resin concentration), the highest dry cell weight (14.97 g/L) and total acetate adsorption (5.45 g/L) were attained. Further increment in sedimented bed height was not possible as the resin was expanded to the top of the column and overflowed. In addition, the optimum superficial linear velocity was determined to be 900 cm/h (degree of expansion of 2), where only a small circular movement of particles but not dispersed was observed. The resin was found to be packed tightly against the upper adaptor net following a bed expansion, when the viscosity was high (113.9 mPa.s), resulting in blockage of column. However, the viscosity of the fermentation broth was determined to be well below that point (15.75 mPas), and hence no problem was posed.

Meanwhile, for a dispersed resin system, the mixing time was found to increase at low agitation speeds (100–200 rpm). In addition, at high agitation speeds (>400 rpm), the mixing time in the integrated STB/EBA system was not significantly different compared to without resin (control) and with dispersed resin system. The performances of three systems: resin-free, dispersed system and STB-EBA integrated system in the fermentation of the PrIFN-α2b-producing recombinant *E. coli* were summarised in **Table 3.** In comparison to without resin and dispersed systems, STB-EBA system exhibited the highest PrIFN-α2b concentration, specific yield and volumetric productivity, shorter growth cycle and time to achieve the maximum PrIFN-α2b concentration. Besides, the simplified clarification of the culture broth (afforded by the use of EBA) reduced the overall length of downstream processing. Notably, with the integrated system, the PrIFN-α2b production was improved by 3-fold and 1.4-fold over that achieved in the resin-free and dispersed system, respectively.

Other reports on the implementation of such STB-EBA set-up can also be found in the literature, albeit with the use of different types of resin [23–26]. Apart from bioreactor, the external unit EBA column has also been incorporated with extraction tank for recovering active compounds from herbs [27, 28].

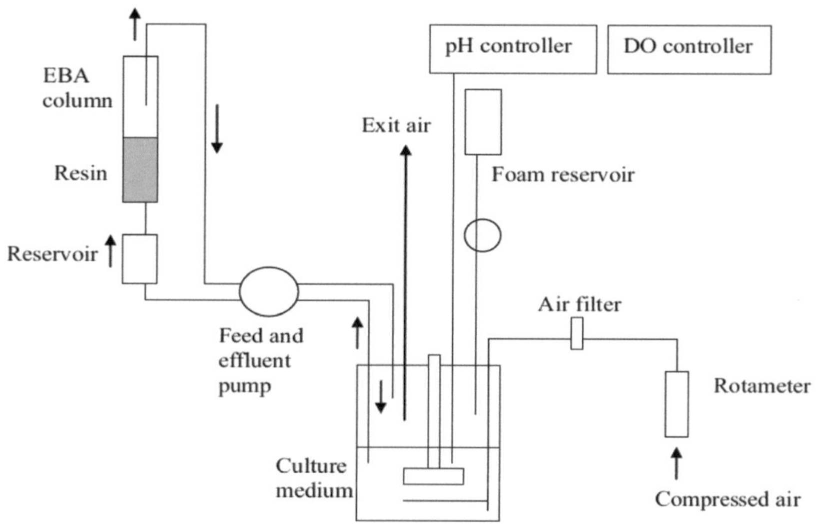

Figure 5. The STB-expanded bed adsorption column design for efficient removal of acetate during the production of PrIFN-α2b by *E. coli* [22].

System.	Maximum cell concentration (g/L)	Time to achieve maximum cell concentration (h)	Maximum PrIFN-α2b concentration (μg/L)	Time to reach maximum PrIFN-α2b concentration (h)	Time to reach threshold level of acetate concentration (≈4.0 g/L)
Resin-free (control)	9.57	16	289.5	12	>13
Dispersed system	12.76	12	638.8	12	>13
Integrated STB-EBA	14.97	24	867.4	11	>21

Table 3. Comparative fermentation performance of the three systems: Resin-free, dispersed system and STB-EBA integrated system in the production of PrIFN-α2b by *E. coli* [22].

3. General methodology in the development of ion-exchange resin-based extractive fermentation

Based on the earlier studies on the extractive fermentation using ion-exchange resin, the general methodology in the development of this technique can be systematically outlined as below:

1. Execution of batch fermentation of a particular culture under a predetermined optimum conditions

Time course fermentation profile of the specific strain will need to be constructed, where the concentrations of cell, substrate concentration, product, and by-product throughout the fermentation are included. The critical concentration of product/by-product, where the cell and product inhibition starts to occur can be identified based on the turning points of the curves.

2. Investigation on the effects of by-products/products on the growth of cell and metabolite production

In the preliminary study for determining inhibitory effect, a pure by-product/product is added to a fermentation culture that is operated in a batch mode operation. The concentration of the by-product/product is varied (which generally covers the range of the real-world concentration as determined earlier) and its effects on the cell growth and metabolite production are monitored. Determination of the threshold level, that is, the critical concentration, where the repression on cell growth and production occur is then being done.

3. Selection of suitable ion-exchange resin for *in situ* adsorption of target product/by-product

The main criteria for selecting a suitable ion-exchange resin mainly depend on the adsorption capacity and selectivity of the resin towards the target product/by-product [29, 30]. Thus, preliminary information on the properties of target molecule such as charge and stable pH range should be gained prior to the addition of resin. The general rule of thumb is that the charge of the selected resin must be the opposite to the charge of the target product/by-product. For *in situ* removal of by-product in a culture that contains extracellular target product, when possible, the charge of the by-product and product must also be different to one and another to ensure selectivity of adsorption. In cases where the charge of by-product and extracellular target product is similar, a resin with higher specificity and affinity towards by-product must be selected. Generally, weak ion-exchangers display a pH-dependent function, while strong ion-exchangers function over a wide pH range. Weak anion-exchange resins are highly ionised only at pH below 7, while weak cation-exchangers start to lose their ionisation below pH 6. Although strong ion-exchangers are generally more powerful as they can cover a wider pH, but their selectivity may not be necessarily higher than weak ion-exchangers. Essentially, when selecting a resin, it is always helpful to refer to the manufacturer's instructions for the pH range of potential resins.

The selection of suitable resin should also be made based on the physical properties of the resin, that is, size and density. The particle size of resin also influences the performance of adsorption, especially when operating in a packed-bed column mode. Smaller size resin particle generally afforded higher resolution in separation (during the elution stage, where packed-bed mode is always preferred to minimise the use of buffer) but will not tolerate high flow rate, which otherwise could results in high backpressure. Conversely, larger size resin particle permits high flow rate but the resolution is relatively lower than the small size resin. Furthermore, in a special system like EBA system, the density of resin also has a significant effect on the performance of adsorption, as it influences the stability of bed expansion. In EBA operation, a larger diameter resin is also preferred as it enables the use of large pore column adapter screen. It is worth to note that the optimal choice of adsorption bead size and density, together with column hardware and operating conditions, ensures maintenance of bed expansion without loss of adsorbent in the column effluent [31].

For integrated packed-bed column-bioreactor system, the *in situ* recovery of product/by-product basically is not a straightforward process and can be rather challenging, considering the viscosity of cell-laden fermentation broth. Therefore, in this system, the selection of a suitable resin size is even more crucial. A bigger resin is generally more tolerable towards a highly viscous fermentation broth in mitigating pressure build-up.

With that, irrespective of any resin application (dispersed, external, or internal systems), initial screening of resin for the adsorption of a specific product/by-product is very crucial, which must be examined on a case-by-case basis, especially when the product and by-product are closely related in terms of charge properties.

Batch sorption experiment is usually being carried out at the first place to select a suitable ion-exchange resin. This is implemented by adding potential ion-exchange resin(s) (with varied concentration) to cell-laden fermentation broth (as cell could significantly influence the adsorption behaviour of adsorption). Experimental data (i.e. concentration at equilibrium and adsorption capacity) are then fitted to an adsorption isotherm such as Freundlich, Langmuir, Brunauer-Emmett-Teller (BET), etc. The Langmuir isotherm is the most commonly used model for describing adsorption isotherm, which describes a mono-layer adsorption with energetically identical sorption sites and without mutual interactions between the adsorbed molecules [32]. Two adsorption parameters, that is, maximum specific uptake capacity (q_m) and dissociated constant (K_d) can then be calculated using a graphical analysis, based on the selected adsorption isotherm model. The selected resin should generally yields high q_m (high binding capacity) and low value of K_d (exhibiting high affinity towards target molecule). In addition, selection should also be made with regard to the desorption efficiency of resin. Some resins might exhibit high binding capacity towards target product/by-product but somehow give low desorption efficiency.

Ideally, apart from the target product/by-product, the resin must show the minimum interaction with all the nutrients in the medium. Nevertheless, during a fermentation, minimum nutrient(s) removal might be tolerated provided that high degree of yield and productivity is achieved. Therefore, a preliminary experiment on *in situ* addition of resin into fermentation culture is very important. Analysis on the content of nutrient ions such as Cl^-, SO_4^{2-}, and PO_4^{2-} and so on in the cultivation medium during the preliminary fermentation (with added resin) may also be done to examine the selectivity of the resin. The information on the co-adsorption of nutrient(s) (if any) by the used resin is useful to provide a more accurate interpretation on the effects of resin on cell growth and metabolite production. Besides, the toxicity effect of the selected resin on cell growth can also be examined through the preliminary experiment. If these factors are found to significantly affecting the fermentation performance, re-selection of resin must be made.

In addition, examination on cell morphology during the fermentation following the addition of resin is also important. The probability of the unwanted cell-resin adsorption and shear effect, which could negatively affect the fermentation performance, can be scrutinised here.

4. Optimisation of adsorption parameters and operating variables

The optimum concentration of ion-exchange resin should be determined with regard to fermentation performance. The general information on the binding capacity of the potential resin (which usually can be obtained from the manufacturer's data), in conjunction with the estimated product/by-product concentration (as determined experimentally earlier in batch fermentation) will help in the rough estimation of the working concentration range. The typical trend observed is improvement in cell growth and metabolite production, when the resin concentration is raised; further, addition of resin will, however, result in suppression of cell growth and metabolite production.

Next, the time of resin addition (in optimum load as determined before), that is, during the inoculation stage or at later stage of fermentation also needs to be determined. Typically, resin addition is done during the inoculation stage for dispersed/internal system; while for external resin system, a resin-loaded column is directly connected to the bioreactor when fermentation is started. Additionally, for external resin system, optimisation of flow rate of medium through the resin column should be carried out. The ideal flow rate is the fastest flow rate that still afforded high adsorption capacity, similar to an ion-exchange chromatography operation. Effects of agitation speed on the culture, in the presence of resin in the culture or when the bioreactor is connected to external resin column, in terms of mixing, shear effect (more significant in dispersed resin system), and DOT may also need to be further optimised.

4. Conclusion

Extractive fermentation using ion-exchange resin for removal of product/by-product of fermentation is mainly employed to mitigate the problem of feedback inhibition. In addition, for *in situ* product removal, additional advantages of shortening downstream processing steps improve process economics and productivity, and minimising product lost and degradation can be achieved.

Research on the selectivity of adsorption or even desorption is very important especially for high-value biologics. However, so far, most of the earlier studies have not focused on the purity aspect of the desorbed products. In addition, it was highlighted in many earlier works that at high resin concentrations, reduction in fermentation performances (i.e. cell growth and metabolite production) was encountered. The main research question: 'what is/are the primary cause(s) of the performance decline - removal of like-charge nutrient component(s), shear effect on cell/product, or reduced mixing efficiency in the presence of high resin load?' however remains. Detailed study on the interaction of resin with medium components may be carried out to gain further insight on this.

In general, the development of new ion-exchange technologies is driven by the needs for having a higher capacity and selective adsorbent. The use of new resins, for example, resins with alternate support matrices, that are able to operate at higher conductivities medium (dilution/diafiltration prior to adsorption will not be required), single-use/disposable resins (to overcome the hassles of cleaning-in-place (CIP), validation, and to improve overall process economic), and so on in the extractive fermentation technique should also be explored.

In addition, the feasibility of combining ion-exchange resin-based adsorptive techniques with other separation technique such as electric fields- and membrane-based technique may also be scrutinised. The use of immobilised cell to protect them from shear stress and/or unwanted interaction with the adsorbent may also be studied. Ultimately, the operation of the ion-exchange resin-based extractive fermentation in continuous mode may deserve special attention. On a separate note, the factor of cost of resin and lifetime of resins should also be taken into account during the selection especially for large-scale operation. Finally, considering that there are abundant of available experimental findings on specific products/by-products-resins adsorption, the data should be generalised to assist the selection of resin in future.

Conflict of interest

The authors declare no financial or commercial conflict of interest.

Notations

ABE	acetone-butanol-ethanol
BET	Brunauer-Emmett-Teller
BP	by-products
CIP	cleaning-in-place
DOT	dissolved oxygen tension
EBA	expanded bed adsorption
hEGF	human epidermal growth factor
K_d	dissociated constant
PrIFN-α2b	periplasmic human interferon-alpha2b
q_m	maximum specific uptake capacity
RSM	response surface methodology
SEM	scanning electron microscope
STB	stirred tank bioreactor
TEM	transmission electron microscope

Author details

Fadzlie Wong Faizal Wong[1,2], Murni Halim[1,2] and Arbakariya B. Ariff[1,2*]

*Address all correspondence to: arbarif@upm.edu.my

1 Department of Bioprocess Technology, Faculty of Biotechnology and Biomolecular Sciences, Universiti Putra Malaysia, Selangor, Malaysia

2 Bioprocessing and Biomanufacturing Research Centre, Faculty of Biotechnology and Biomolecular Sciences, Universiti Putra Malaysia, Selangor, Malaysia

References

[1] Tan JS, Ramanan RN, Ling TC, Shuhaimi M, Ariff AB. Enhanced production of periplasmic interferon alpha-2b by *Escherichia coli* using ion-exchange resin for in situ removal of acetate in the culture. Biochemical Engineering Journal. 2011;**58-59**:124-132. DOI: 10.1016/j.bej.2011.08.018

[2] Vane LM. A review of pervaporation for product recovery from biomass fermentation processes. Journal of Chemical Technology and Biotechnology. 2005;**80**(6):603-629. DOI: 10.1002/jctb.1265

[3] Yang S-T, Huang H, Tay A, Qin W, Guzman LD, ECS N. Extractive fermentation for the production of carboxylic acids. In: Yang S-T, editor. Bioprocessing for Value-Added Products from Renewable Resources. Amsterdam: Elsevier; 2007. pp. 421-446. DOI: 10.1016/B978-044452114-9/50017-7

[4] Herrero AA. End-product inhibition in anaerobic fermentations. Trends in Biotechnology. 1983;**1**(2):49-53. DOI: 10.1016/0167-7799(83)90069-0

[5] Kim C-H, Kim S-W, Hong S-I. An integrated fermentation–separation process for the production of red pigment by *Serratia* sp. KH-95. Process Biochemistry. 1999;**35**:485-490. DOI: 10.1016/S0032-9592(99)00091-6

[6] Liu B, Hui J, Cheng YQ, Zhang X. Extractive fermentation for enhanced production of thailandepsin A from *Burkholderia thailandensis* E264 using polyaromatic adsorbent resin Diaion HP-20. Journal of Industrial Microbiology & Biotechnology. 2012;**39**:767-776. DOI: 10.1007/s10295-011-1073-x

[7] Hecke WV, Kaur G, DeWever H. Advances in in-situ product recovery (ISPR) in whole cell biotechnology during the last decade. Biotechnology Advances. 2014;**32**:1245-1255. DOI: 10.1016/j.biotechadv.2014.07.003

[8] Schügerl K, Hubbuch J. Integrated bioprocesses. Current Opinion in Microbiology. 2005;**8**:294-300. DOI: 10.1016/j.mib.2005.01.002

[9] Buque-Taboada EM, Straathof AJJ, Heijnen JJ, Wielen LAMVD. In situ product recovery (ISPR) by crystallization: Basic principles, design, and potential applications in

whole-cell biocatalysis. Applied Microbiology and Biotechnology. 2006;**71**(1):1-12. DOI: 10.1007/s00253-006-0378-6

[10] Carstensen F, Apel A, Wessling M. In situ product recovery: Submerged membranes vs. external loop membranes. Journal of Membrane Science. 2012;**394-395**:1-36. DOI: 10.1016/j.memsci.2011.11.029

[11] Freeman A, Woodley JM, Lilly MD. In situ product removal as a tool for bioprocessing. Bio/Technology. 1993;**11**:1007-1012. DOI: 10.1038/nbt0993-1007

[12] Hazwani-Oslan SN, Tan JS, Saad MZ, Halim M, Ariff AB. Improved cultivation of gdhA derivative *Pasteurella multocida* B:2 for high density of viable cells through in situ ammonium removal using cation-exchange resin for use as animal vaccine. Process Biochemistry. 2017;**56**:1-7. DOI: 10.1016/j.procbio.2017.02.016

[13] Kadam PC, Boone DR. Influence of pH on ammonia accumulation and toxicity in halophilic, methylotrophic methanogens. Applied and Environmental Microbiology. 1996;**62**:4486-4492. PMCID: PMC1389003

[14] Wee YJ, Kim JN, Ryu HW. Biotechnological production of lactic acid and its recent applications. Food Technology and Biotechnology. 2006;**44**:163-172

[15] Cui S, Zhao J, Zhang H, Chen W. High-density culture of lactobacillus plantarum coupled with a lactic acid removal system with anion exchange resins. Biochemical Engineering Journal. 2016;**115**:80-84. DOI: 10.1016/j.bej.2016.08.005

[16] Othman M, Ariff AB, Wasoh H, Kapri MR, Halim M. Strategies for improving production performance of probiotic *Pediococcus acidilactici* viable cell by overcoming lactic acid inhibition. AMB Express. 2017;**7**:215. DOI: 10.1186/s13568-017-0519-6

[17] Chen X, Cen P, Chen J. Enhanced production of human epidermal growth factor by a recombinant *Escherichia coli* integrated with in situ exchange of acetic acid by macroporous ion-exchange resin. Journal of Bioscience and Bioengineering. 2005;**100**(5):579-581. DOI: 10.1263/jbb.100.579

[18] Lee JC, Park HR, Park DJ, Lee HB, Kim YB, Kim CJ. Improved production of teicoplanin using adsorbent resin in fermentations. Letters in Applied Microbiology. 2003;**37**:196-200. DOI: 10.1046/j.1472-765X.2003.01374.x

[19] Jia B, Jin ZH, Lei YL, Mei LH, Li NH. Improved production of pristinamycin coupled with an adsorbent resin in fermentation by *Streptomyces pristinaespiralis*. Biotechnology Letters. 2006;**28**:1811-1815. DOI: 10.1007/s10529-006-9157-9

[20] Bae J, Moon H, Oh KK, Kim CH, Sil Lee D, Kim SW, Hong SI. A novel bioreactor with an internal adsorbent for integrated fermentation and recovery of prodigiosin-like pigment produced from *Serratia* sp. KH-95. Biotechnology Letters. 2001;**23**:1315-1319. DOI: 10.1023/A:1010573427080

[21] Wang J, Wen X, Zhou D. Production of citric acid from molasses integrated with in-situ product separation by ion-exchange resin adsorption. Bioresource Technology. 2000;**75**:231-234. DOI: 10.1016/S0960-8524(00)00067-5

[22] Tan JS, Ling TC, Mustafa S, Tam YJ, Ramanan RN, Ariff AB. An integrated bioreactor-expanded bed adsorption system for the removal of acetate to enhance the production of alpha-interferon-2b by *Escherichia coli*. Process Biochemistry. 2013;**48**:551-558. DOI: 10.1016/j.procbio.2013.02.024

[23] Ohashi R, Otero JM, Chwistek A, Yamato I, Hamel JP. On-line purification of monoclonal antibodies using an integrated stirred-tank reactor/expanded-bed adsorption system. Biotechnology Progress. 2002;**18**:1292-1300. DOI: 10.1021/bp025543m

[24] Li Q, Wang D, Hu G, Xing J, Su Z. Integrated bioprocess for high-efficiency production of succinic acid in an expanded-bed adsorption system. Biochemical Engineering Journal. 2011;**56**:150-157. DOI: 10.1016/j.bej.2011.06.001

[25] Wang P, Wang Y, Liu Y, Shi H, Su Z. Novel in situ product removal technique for simultaneous production of propionic acid and vitamin B12 by expanded bed adsorption bioreactor. Bioresource Technology. 2012;**104**:652-659. DOI: 10.1016/j.biortech.2011.10.047

[26] Tripathi A, Kumar A. Integrated approach for β-glucosidase purification from non-clarified crude homogenate using macroporous cryogel matrix. Separation Science and Technology. 2013;**48**(16):2410-2417. DOI: 10.1080/01496395.2013.807833

[27] Zhang M, Hu P, Liang Q, Yang H, Liu Q, Wang Y, Luo G. Direct process integration of extraction and expanded bed adsorption in the recovery of crocetin derivatives from Fructus gardenia. Journal of Chromatography B. 2007;**858**:220-226. DOI: 10.1016/j.jchromb.2007.08.041

[28] Zhang M, Yang H, Chen X, Zhou Y, Zhang H, Wang Y, Hu P. In-situ extraction and separation of salvianolic acid B from Salvia miltiorrhiza Bunge by integrated expanded bed adsorption. Separation and Purification Technology. 2011;**80**(3):677-682. DOI: 10.1016/j.seppur.2011.06.039

[29] Pharmaceutical Technology. Advances in Ion-Exchange Chromatography Leading to Increased Efficiencies and Reduced Costs [Internet]. 2013. Available from: http://www.pharmtech.com/advances-ion-exchange-chromatography-leading-increased-efficiencies-and-reduced-costs?pageID=1 [Accessed: 2018-01-25]

[30] Otman M, Ariff AB, Rios-Solis L, Halim M. Extractive fermentation of lactic acid in lactic acid bacteria cultivation: A review. Frontiers in Microbiology. 2017;**8**:2285. DOI: 10.3389/fmicb.2017.02285

[31] Glover DJ, Humphreys DP. Antibody fragments: Production, purification and formatting for therapeutic applications. In: Subramanian G, editor. Antibodies. Volume 1: Production and Purification. New York, Springer, Science+Business Media New York; 2004. pp. 75-100. DOI: 10.1007/978-1-4419-8875-1

[32] Liu J, Su Y, Li Q, Yue Q, Gao B. Preparation of wheat straw based superabsorbent resins and their applications as adsorbents for ammonium and phosphate removal. Bioresource Technology. 2013;**143**:32-39. DOI: 10.1016/j.biortech.2013.05.100

Chapter 4

Hydroxide Transport in Anion-Exchange Membranes for Alkaline Fuel Cells

Sergio Castañeda Ramírez and
Rafael Ribadeneira Paz

Additional information is available at the end of the chapter

http://dx.doi.org/10.5772/intechopen.77148

Abstract

In this chapter are described the characteristics of transport of hydroxide ions through hydrated polymeric materials with potential application in alkaline fuel cells are described. First, it is made a brief description of anion-exchange membrane fuel cells (AEMFCs), their evolution and key characteristics. Then, this chapter presents a detailed classification of the different types of polymers that have been proposed for AEMFCs and their state of development. After that, mechanisms involved in the transport of hydroxide ions through hydrated anion-exchange membranes are described and discussed, making emphasis in the theoretical approaches applied for their study and their implementation and representability in global transport models. In the final section, it is summarized the key features of the chapter and is made a brief discussion about challenges and future work required for the consolidation of this promising technology.

Keywords: fuel cell, anion-exchange, polymer, hydroxide ion, transport phenomena

1. Introduction

A fuel cell is a device that converts the chemical energy contained in a fuel directly and efficiently (between 60 and 80% at normal operating conditions) into electrical energy. This is achieved by the coupling of reduction/oxidation reactions between oxygen and fuel fed respectively to the cathode and anode of the cell, having as by-products water and CO_2 in low amounts when a hydrocarbon (such as methanol or ethanol) is the fuel [1, 2].

The general scheme of an alkaline fuel cell (AFC) is shown in **Figure 1**. In this device, fuel is oxidized at the anode and combined with hydroxide ions transported from the cathode

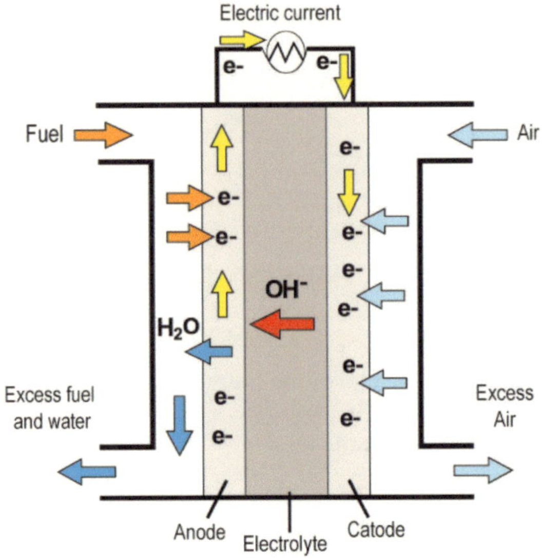

Figure 1. General scheme of an AFC.

through the electrolyte to produce water. Part of water leaves the cell while the rest moves through the electrolyte to the cathode. There, it is combined with oxygen and electrons coming from the anode through an external circuit to produce hydroxide ions that are then transported through the electrolyte to the anode [3–5]. The electrochemical reactions involved if the fuel is hydrogen are:

$$\text{Anode}: 2H_2 + 4OH^- \rightarrow 4H_2O + 4e^- \tag{1}$$

$$\text{Cathode}: O_2 + 2H_2O + 4e^- \rightarrow 4OH^- \tag{2}$$

$$\text{Overall}: 2H_2 + O_2 \rightarrow 2H_2O \tag{3}$$

The most known types of AFCs use liquid solutions as electrolytes because of their high ionic conductivity [5, 6]. However, they have as disadvantage their short lifetime under normal operating conditions mainly due to the formation of carbonates (CO_3^{2-}) between hydroxides of the electrolyte and CO_2 [7–10] and the permeation of liquid electrolyte to electrodes (weeping) [10]. In consequence, polymer-based membranes for AFCs are developed and implemented [11, 12] following the model of their counterparts, the proton-exchange membrane fuel cells (PEMFCs). With this, the newly born AEMFCs were made suitable for their use in vehicles and batteries for mobile devices because of: (a) the conductive species of the electrolyte are integrated to a rigid polymeric matrix so reactivity and production of carbonates is reduced, (b) there is no weeping because the electrolyte is a solid whose segregation is minimum, and (c) the versatility of the fuel cell is better with respect to implementation and handling using the polymeric electrolyte [3, 6, 13].

Despite this, the polymeric materials used as electrolytes in AEMFCs currently show drawbacks susceptible of research and development in their ionic conductivity and chemical stability. Such characteristics limit the efficiency and durability of the fuel cell respectively; thus, their study is the main focus of research in this area. Most of studies are experimental and focused on the fabrication and characterization of polymeric materials with potential application in AEMFCs. The others are theoretical and seek to represent transport phenomena of hydroxide ions through anion-exchange membranes to understand their characteristics and propose how to improve them. Therefore, this chapter is developed as follows: first, it reviews and discusses the contributions and findings made to date about anion-exchange membranes for AEMFCs from both experimental (Section 2) and theoretical studies (Section 3). Then, it addresses the challenges and work required to consolidate the AEMFCs and make them a feasible energy production technology for the close-future.

2. Experimental studies

The experimental research of anion-exchange membranes for AEMFCs has focused on the development of materials having: (a) good mechanical and thermal properties at the assembly and operation conditions of the fuel cell [14–16], (b) chemical stability especially against the attack of hydroxide ions to the cationic functional groups of the membrane [17–19], (c) high anionic conductivity (>100 mS/cm) [6, 20, 21], (d) electronic isolation [22], (e) appropriate thickness (between 50 and 80 μm) [3], (f) low permeability to fuel [9], and (g) low cost [23, 24]. In order to fulfill them, diverse anion-exchange membranes have been fabricated and studied. These membranes can be classified according to their structure as: (a) heterogeneous, subclassified according to the nature of their components, (b) interpenetrated polymer networks, and (c) homogeneous, subclassified according to their functionalization method [3]. This is schematically shown in **Figure 2**.

2.1. Heterogeneous membranes

A heterogeneous membrane is an anion-exchange material embedded in an inert compound. According to their nature, these membranes are classified into:

2.1.1. *Ion solvating polymers*

They are composed of a polymeric matrix soluble in water that contains electronegative heteroatoms (such as oxygen, nitrogen, sulfurs, chlorides or phosphates), a hydroxide (most of the time potassium hydroxide) and one or more plasticizers. The resultant material has the mechanic properties of the polymeric matrix and the electrochemical properties of the hydroxide [25, 26]. Among the materials used to fabricate ion solvating polymers, there are the polyethylene oxide (PEO), polyvinyl alcohol (PVA), chitosan and polybenzimidazole (PBI). They have good mechanical properties but low ionic conductivity at contact with the fuel cell electrodes (between 0.1 and 1 mS/cm) [27] because they are usually very thick and have high electrical resistance. Moreover, their structure is usually not uniform and has areas of high and low concentrations of exchangeable ionic groups that cause an uneven ionic transport through

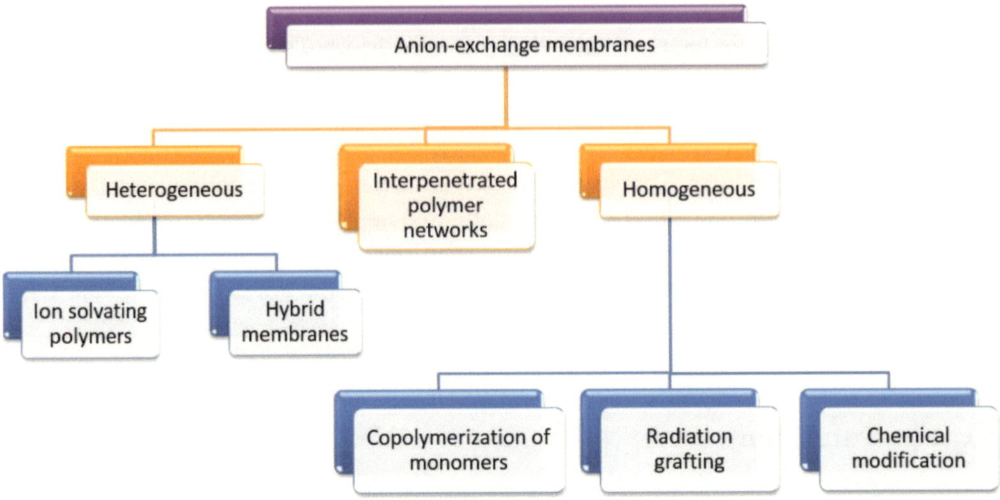

Figure 2. Classification of anion-exchange membranes according to their structure and functionalization method. Reprinted from *J. Memb. Sci.*, 377(1–2), G. Merle, M. Wessling, K. Nijmeijer, Anion exchange membranes for alkaline fuel cells: A review, 1-35, Copyright 2011, with permission from Elsevier.

the material. Additionally, the KOH used as a hydroxide is susceptible to leak out the membrane, resulting in continuous conductivity losses during the operation of the fuel cell [20].

2.1.2. *Hybrid membranes*

They are composed of organic and inorganic segments. Organic segments provide the electrochemical properties and the inorganic (usually silica or siloxane), the mechanical properties. Examples of hybrid membranes are PEO, PVA and polyphenylene oxide (PPO) with silica (SiO_2) and membranes based on PVA with titanium dioxide (TiO_2). Although their mechanical properties are promising due to the incorporation of inorganic components, membranes of this category have the same nonuniformity problem of ion solvating polymers, thus their ionic conductivity are similar or even lower [15].

2.2. Interpenetrated polymer networks

These membranes combine two polymeric materials by crosslinking without promoting the formation of covalent bonds in between. One of the polymers is hydrophobic and has good chemical, mechanical, and thermal properties, while the other is an ionic conductor. Main researches for membranes in this category are related with materials based on PVA, PPO, polyethylene, and polysulfone. They are easy to fabricate, and the range of polymers that can be used is wide. Additionally, they exhibit low electrical resistance and high mechanical strength, chemical stability, durability, and can be produced at a reasonable cost. However, since the constituent materials are not covalently bonded, the conductive polymer slowly diffuses out of the membrane with time, which causes progressive losses of conductivity and ion exchange capacity. Furthermore, although these membranes can reach ionic conductivities higher than heterogeneous membranes (2–5 mS/cm), they are still very low for their use in a fuel cell [3, 14].

2.3. Homogeneous membranes

They are polymers composed of one type of material that is modified to have ion exchange capacity. In these materials, the cationic functional groups (usually quaternary amines) are covalently attached to the polymer backbone to generate ionic sites with an associated mobile counterion [6, 28]. According to how they are functionalized, homogeneous membranes can be classified into:

2.3.1. Copolymerization of monomers

Copolymerization is a process in which 2 or more monomers combine into a single polymeric chain. The most promising membranes produced by this method are based on chloromethylstyrene and divinylbenzene. However, they have as disadvantage the low availability of chloromethylstyrene and high cost of divinylbenzene [3, 28].

2.3.2. Radiation grafting

These membranes are conformed by a hydrophobic polymer backbone to which cationic functional groups are grafted by irradiation, UV, or plasma methods [3]. As the most important materials in this category are the poly(fluorinated ethylene-propylene) (FEP) and poly(ethylene-*co*-tetrafluoroethylene) (ETFE) membranes, both functionalized with polyvinylbenzene (PVB) and trimethylammonium hydroxide groups. They have a very high ionic conductivity with respect to other types of membranes (between 10 and 45 mS/cm) and excellent mechanical properties. However, they are currently impractical for massive production because of the high cost of their constituent components (especially for fluorinated membrane) and the grafting processes [6, 20].

2.3.3. Chemical modification

These membranes are the most researched due to they can reach high ionic conductivities at a lower fabrication cost than membranes produced by copolymerization and radiation grafting. However, chemical modification methods are difficult to control and reproduce. In consequence, the resulting quality of the membranes is highly dependent on the rigor of the fabrication procedure. Some of the anion-exchange membranes of this category are based on polystyrene, PVA, PVB, epichlorohydrin, polyethylene glycol, ethers, aromatic esters, guanidine, polysulfone, and polypropylene [3, 6].

Homogeneous membranes are to date the most efficient and promising electrolytes for AEMFCs because of their superior ionic conductivity and durability with respect to the other membrane categories. However, yet there is no a consolidated material that fulfills the conditions required to achieve an optimal operation in an AEMFC, despite the fact that significant advances have been made in the last years with respect to the quality of their fabrication and properties. For that reason, it is essential not only to improve them from the experimental work, but also with theoretical studies to understand the characteristics of the transport mechanisms of hydroxide ions through anion-exchange membranes and how they influence their ionic conductivity and chemical stability as it is shown below.

3. Theoretical studies

As was introduced in previous sections, a good alternative to design efficient anion-exchange membranes is the implementation of theoretical models to identify, analyze, and complement the experimental findings about transport phenomena within them. However, studies of this type are few, and the characteristics of such transport phenomena are not well known and are still in discussion to date [29]. In order to identify them, studies for transport of hydronium ions (H_3O^+) in hydrated proton-exchange membranes have been taken as a starting point under the premise that the characteristics of ionic transport for both anionic and protonic membranes could be similar (which is not strictly true) [30, 31]. Thus, in **Figure 3** are schematically shown the main transport mechanisms for hydroxide ions through anion-exchange membranes that have been proposed in the literature. These include diffusion, which consist of molecular or *en masse* diffusion, structural diffusion or Grotthuss mechanism and surface site hopping; migration and convection.

3.1. Diffusion

It is defined as the transport of molecules due to a chemical potential gradient of one or more components in the absence of an electric field [32]. Within a hydrated anion-exchange membrane, hydroxide ions can diffuse by three modalities: the traditional *en masse* diffusion, Grotthuss mechanism, and surface site hopping. The first two mechanisms take place at the

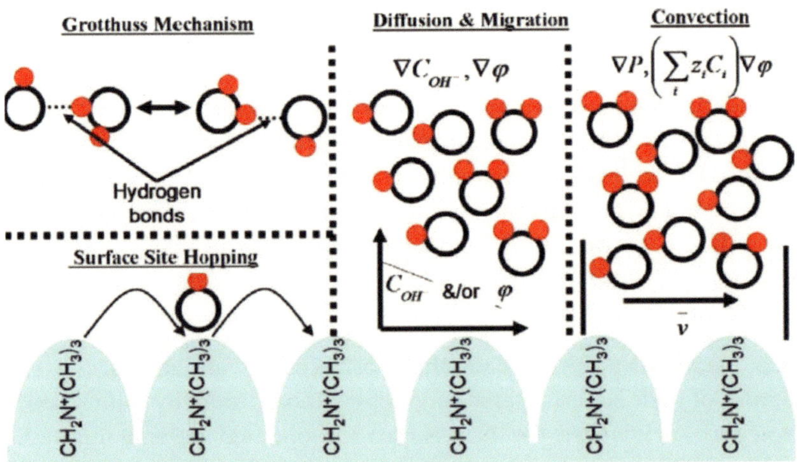

Figure 3. Transport mechanisms proposed for hydroxide ions in anion-exchange membranes. Reprinted with permission from *J. Electrochem. Soc.*, 2010; 157(3): B327. Copyright 2010, The Electrochemical Society.

bulk of water molecules (usually in the middle region of the membrane pores and far from its backbone and side chains), while the latter takes place on the surface of the polymer cationic functional chains [33]. A visualization of the way hydroxide ions can diffuse by these mechanisms is depicted as an electric circuit in **Figure 4**.

In which N_{OH^-} is the total flux of hydroxide ions transported through the membrane, $N_{OH^-}^{Surf}$ is the flux of hydroxide ions transported through the surface of the polymer chains and $N_{OH^-}^{Bulk}$ is the flux of hydroxide ions transported through the region of bulk water. Additionally, $D_{OH^-}^{Surf}$ and $D_{OH^-}^{Grott}$ are the contributions to total diffusion coefficient from surface site hopping and Grotthuss mechanisms, while $D_{OH^-,W}^{M}$ and D_{W,OH^-}^{M} are respectively the contributions from *en masse* diffusion of hydroxide ions and water molecules.

Contributions of each mechanism to total diffusion are strongly influenced by the nanostructure and water content of the anion-exchange membrane [33, 34]. On the one hand, at low hydration levels, pores of the polymer are narrow and cationic side chains are very close to each other. In consequence, solvation of water molecules and interaction between them and hydroxide ions via hydrogen bonding are very low with respect to the electrostatic forces exerted by the cationic chains on the latters. Therefore, transport of hydroxide ions is more likely to take place on the surface of the polymeric chains by surface site hopping and molecular diffusion at low rates. In addition, ionic conductivity of the anion-exchange membrane will be very low. On the other hand, at high hydration levels, the pores of the polymer swell and give place to wide continuous channels where regions of bulk water can be formed. Cationic functional chains will be more separated, thus their influence on hydroxide ions is reduced and dissociation from solvation of water molecules will be more likely. Therefore, transport of hydroxide ions takes place mostly by charge defect transfer (Grotthuss mechanism) in the bulk water region at high rates: the ionic conductivity of the anion-exchange membrane under that condition will reach its highest values [34].

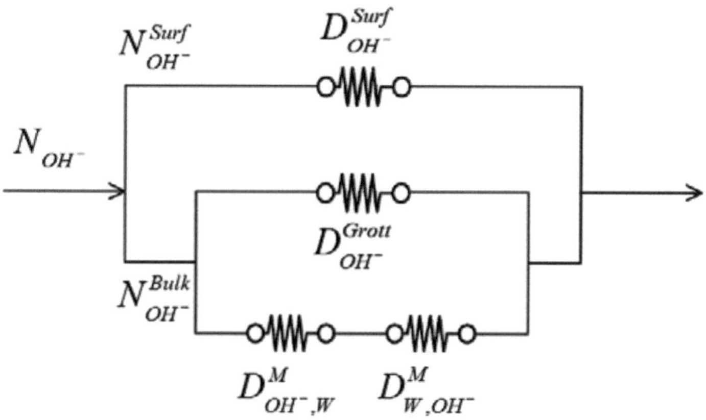

Figure 4. Electric circuit analogy of the transport mechanisms by which hydroxide ions can diffuse through anion exchange membranes. Reprinted with permission from *J. Electrochem. Soc.*, 2005; 152(3): E123. Copyright 2005, The Electrochemical Society.

In order to gain deeper understanding of the mentioned diffusion mechanisms, a general description for each one and their implementation in mathematical models is given in the next subsections.

3.1.1. en masse *diffusion*

The molecular movement of hydroxide ions due to concentration (or activity) gradients can be described according to Fick's law:

$$N^{M}_{OH^{-}} = -D^{M}_{OH^{-},W}\nabla c_{OH^{-}} \tag{4}$$

in which $N^{M}_{OH^{-}}$ is the flux of hydroxide ions due to *en mass* diffusion and $c_{OH^{-}}$ their corresponding concentration. However, since water molecules are also diffusing by their own gradient and there are also frictional interactions with the membrane, it is more rigorous to use the multicomponent Stefan-Maxwell equation to take into account those effects in the mass diffusion of hydroxide ions [29, 32]:

$$\nabla x_i = \sum_{j\neq i}^{n}\frac{x_iN_j - x_jN_i}{c_T D^{M}_{i,j}} - \frac{N_i}{c_T D^{M}_{i,AEM}} \tag{5}$$

in which x_i is the mole fraction of specie i, c_T the total concentration of all species, $D^{M}_{i,j}$ the mass diffusion coefficient between species i and j and $D^{M}_{i,AEM}$ the mass diffusion coefficient between specie i and the membrane structure. Expressing Eq. (5) for hydroxide ions in a hydrated membrane, it becomes:

$$\nabla x_{OH^{-}} = \frac{x_{OH^{-}}N_W - x_W N_{OH^{-}}}{c_T D^{M}_{OH^{-},W}} - \frac{N_{OH^{-}}}{c_T D^{M}_{OH^{-},AEM}} \tag{6}$$

The first term of the right side of Eq. (6) corresponds to the interactions between water molecules and hydroxide ions and the other represents the frictional effects of the membrane. In addition, to account for Knudsen diffusion, binary diffusion coefficients can be expressed as a parallel resistance with the following form [32]:

$$D^{M,eff}_{i,j} = \frac{1}{\frac{1}{D^{M}_{i,j}} + \frac{1}{D_{Ki}}} \tag{7}$$

The Knudsen diffusion coefficient (D_{Ki}) can be approximated from the kinetic theory of gases [32]:

$$D_{Ki} = \frac{d}{3}\sqrt{\frac{8RT}{\pi M_i}} \tag{8}$$

in which d is the pore diameter of the membrane, R the ideal gas constant, T the temperature, and M_i the molecular weight of specie i. Alternatively, effective diffusion coefficient of Eq. (7) can be expressed in terms of porosity (ε) and tortuosity (τ) by the following relation [35]:

$$D_{i,j}^{M,eff} = \frac{\varepsilon}{\tau} D_{i,j}^{M} \tag{9}$$

or by using the Bruggeman percolation model, initially used for hydronium ions in proton-exchange membranes but later extended to hydroxide ions in anion-exchange membranes [29]:

$$D_{i,j}^{M,eff} = (v_w - v_{w,o})^q D_{i,j}^{M} \tag{10}$$

in which v_w is the volume fraction in the membrane, $v_{w,o}$ the volume fraction of water at percolation limit (minimum water content required to allow transport of hydroxide ions through the membrane) and q the Bruggeman constant.

Binary diffusion coefficients ($D_{i,j}^{M}$ and $D_{i,AEM}^{M}$) must be obtained either from experimental measurements or empirical correlations. By including effective diffusion coefficients in Eq. (6), it gets the following form:

$$\nabla x_{OH^-} = \frac{x_{OH^-} N_W - x_W N_{OH^-}}{c_T D_{OH^-,W}^{M,eff}} - \frac{N_{OH^-}}{c_T D_{OH^-,AEM}^{M,eff}} \tag{11}$$

In order to obtain the flux and mole fraction profiles for hydroxide ions from Eq. (11), it must be solved simultaneously with the Stefan-Maxwell equation for water:

$$\nabla x_W = \frac{x_W N_{OH^-} - x_{OH^-} N_W}{c_T D_{W,OH^-}^{M,eff}} - \frac{N_W}{c_T D_{W,AEM}^{M,eff}} \tag{12}$$

In addition, appropriate boundary conditions must be established. These can be obtained by coupling the transport model of Eqs. (11) and (12) with a global model for the AEMFC.

3.1.2. Grotthuss mechanism

Also known as structural diffusion and proton hopping, it is a transport mechanism by which a protonic excess (hydronium (H_3O^+) for instance) or defect (hydroxide (OH^-) for instance) of an ionic specie diffuses through the hydrogen bond network of water molecules by means of a reactive process caused by fluctuations in the coordination bonds between ions and water that involve the formation and cleavage of hydrogen bonds [36–39].

Currently, the exact description of Grotthuss mechanism for hydroxide ions is still in discussion, and different theories have been proposed. Among them, the dynamic hypercoordination theory is considered the most accurate description to date. However, it is postulated for pure aqueous medium and has not been extended to consider the presence of an anion-exchange membrane.

The steps involved in the Grotthuss mechanism according to dynamic hypercoordination theory are shown in **Figure 5**. It is based on the presolvation concept, which establishes that species with charge defects must be first solvated by water molecules to perform the charge defect transfer [38, 40]. For just water, its molecules form tetrahedral hypercoordinated complexes with adjacent

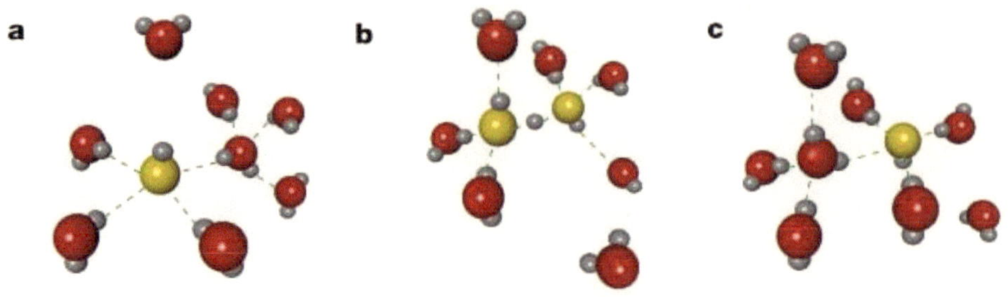

Figure 5. Grotthuss mechanism for hydroxide ions in pure water is according to the dynamic hypercoordination theory. Colors code: oxygen of water molecules is in red, oxygen of molecules with charge defects is in yellow, and hydrogen is in gray. Reprinted by permission from Springer Customer Service Centre GmbH: Springer Nature, The nature and transport mechanism of hydrated hydroxide ions in aqueous solution, M.E. Tuckerman, D. Marx, Copyright 2002.

molecules by donating and receiving two hydrogen bonds, respectively [41, 42] as shown in **Figure 6**. According to the hypercoordination theory, when a hydroxide ion goes into the water network, it adopts a square-planar topology (**Figure 5a**) in which its oxygen atom accepts four hydrogen bonds from neighboring water molecules (forming the anion $H_9O_5^-$). In addition, the hydrogen atom of the hydroxide ion is delocalized around the oxygen atom and stays without establishing coordination bonds. To carry out the charge defect transfer, hydroxide ion must first reduce its coordination number by breaking one of the hydrogen bonds received from a water molecule and then establishing bonding between its hydrogen atom and another nearby water molecule. This allows the ion to take the topology of a fully coordinated water molecule, promoting the transfer of the anionic defect to an adjacent molecule in a process in which the complex $H_3O_2^-$ is temporarily formed (**Figure 5b**). When the transfer is finished, the receiving molecule rearranges to take the preferential square-planar configuration of a hydroxide ion, thereby completing the transport process (**Figure 5c**) [36–38, 40].

It is considered that Grotthuss mechanism has a predominant contribution to hydroxide mobility through hydrated membranes according to: (a) experimental studies about hydroxide mobility in pure aqueous medium [43, 44], (b) theoretical studies about hydroxide mobility in

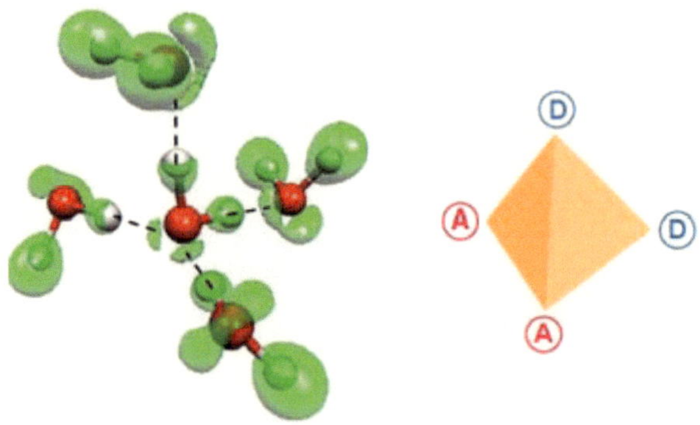

Figure 6. Left: connectivity of a water molecule with four adjacent water molecules. Electronic clouds are indicated in green. Right: tetrahedral representation of the water molecule indicating the donated (D) and accepted (A) hydrogen bonds. From D. Marx: Throwing Tetrahedral Dice. Science. 2004; 303: 634–636. Reprinted with permission from AAAS.

pure aqueous medium with *ab initio* Molecular Dynamics (AIMD) [37, 38], and (c) analogous studies for PEMFCs [33, 45–48].

3.1.3. Surface site hopping

It involves the movement of hydroxide ions by means of successive hops from one side chain of the membrane to another due to strong electrostatic attractive forces exerted by the cationic functional groups on the ions [3, 29, 33]. This process takes place as follows: first, a hydroxide ion attached to a cationic functional group is solvated and dissociated by water molecules. After that, an adjacent side chain attracts the solvated ion to its surface, then the process repeats. This results in a net displacement of the hydroxide ion through the membrane equal to the distance between the two cationic side chains.

Although this mechanism is more likely at low water contents of the membrane, it is considered a secondary process because of the strong interactions between water molecules in the system and the hydrophilic cationic functional groups, which act as a barrier for the hydroxide ions to interact and reach the surface of the latters. This reduces the possibility of this mechanism to take place over the others [29].

Both Grotthuss mechanism and surface site hopping take place at atomic length and time scale and can only be effectively studied through quantum physic techniques like AIMD because of the nature of this phenomenon. At the macroscale, their contributions to total diffusion in mathematical models can be accounted in following way: Grotthuss mechanism, which takes place at the bulk of water molecules, can be accounted by extending the expression for the effective mass diffusion of hydroxide ions in water (Eq. (7)) as:

$$D_{OH^-,W}^{eff} = \frac{1}{\frac{1}{D_{OH^-,W}^{M,eff}} + \frac{1}{D_{OH^-}^{Grott}}} \tag{13}$$

Surface site hopping can be accounted by applying empirical corrections to the influence of the membrane structure on hydroxide mobility (that is $D_{OH^-,AEM}^{M,eff}$) to include not only frictional effects but also surface phenomena [29]. By applying Eq. (13) and an appropriate correction to $D_{OH^-,AEM}^{M,eff}$into Eq. (11), one gets:

$$\nabla x_{OH^-} = \frac{x_{OH^-}N_W - x_W N_{OH^-}}{c_T D_{OH^-,W}^{eff}} - \frac{N_{OH^-}}{c_T D_{OH^-,AEM}^{eff}} \tag{14}$$

in which $D_{OH^-,AEM}^{eff}$ is the effective diffusion coefficient between hydroxide ions and membrane that accounts for surface site hopping. Eq. (14) combined with Eq. (13) are rigorous expressions that take into account all the diffusive phenomena that hydroxide ions can undergo. However, its full application is very limited because Grotthuss mechanism and surface site hopping have not been formally characterized for anion-exchange membranes, and thus, there are no accurate correlations to represent them (i.e., expressions for $D_{OH^-}^{Grott}$ to account for Grotthuss mechanism and either $D_{OH^-,AEM}^{eff}$ or $D_{OH^-}^{Surf}$ in **Figure 4** to account for surface site hopping) like for proton-exchange membranes (see for instance, the research of Choi et al. [33]). For that reason, current transport models approximate $D^{eff}_{OH^-,W}$ by combining the value of the binary diffusion coefficient of hydroxide ions in pure liquid water at 25°C (5.3×10^{-9} m^2/s) with empirical correlations to take into account effects of temperature, pressure and water content, which in practice works well to obtain accurate solutions to the transport models but screening how exactly diffusion of hydroxide ions is taking place and how it changes with temperature, pressure, and water content.

3.2. Convection

Convective transport can take place mainly due to a pressure gradient between the boundaries of the anode and cathode diffusion layers of the fuel cell and the membrane. It is also due to electro-osmotic drag, in which a flux of water molecules is induced by the motion of hydroxide ions in the absence of concentration gradients (i.e., by electric potential gradients) [3, 20, 29].

In electrochemical systems, a convection velocity can be defined by means of the Schogel's equation [29]:

$$v_{conv} = \frac{B_o}{\eta}\left[\nabla P + \left(\sum_i^n z_i c_i\right) F\nabla\phi\right] \tag{15}$$

in which B_o is the d' Arcy hydraulic permeability, η the dynamic viscosity, F the Faraday constant, z_i the charge number of specie i and ϕ the electric potential. The terms in the brackets at right side of Eq. (15) are respectively the contributions to convection by pressure gradients and electro-osmotic drag. It is important to point out that Schoegel's equation assumes that charged species in the radial direction of the membrane pores are uniformly distributed, thus potential gradients are neglected. This is indeed the case for exchange membranes in which pore sizes are small so there are not considerable localized variations in the radial direction [29].

If transport by convection and diffusion are taking place simultaneously, Eq. (15) can be combined with Eq. (5):

$$\nabla x_i = \sum_{j\neq i}^n \frac{x_i N_j - x_j N_i}{c_T D^{eff}_{i,j}} - \frac{N_i}{c_T D^{eff}_{i,AEM}} - \frac{B_o}{\eta D^{eff}_{i,AEM}}\left[\nabla P + \left(\sum_i^n z_i c_i\right) F\nabla\phi\right] \tag{16}$$

in which binary diffusion coefficient was replaced by effective diffusion coefficients to include the effect of membrane structure and all the possible diffusion mechanisms affecting specie i (as was described in Section 3.1). In addition, convection term appears divided by $D^{eff}_{i,AEM}$ to take into account any frictional effect of the membrane on this mechanism. Applying the abovementioned definitions into Eqs. (11) and (12), the following set of equations is obtained:

$$\nabla x_{OH^-} = \frac{x_{OH^-}N_W - x_W N_{OH^-}}{c_T D^{eff}_{OH^-,W}} - \frac{N_{OH^-}}{c_T D^{eff}_{OH^-,AEM}} - \frac{B_o}{\eta D^{eff}_{OH^-,AEM}}\left[\nabla P + c_{OH^-}F\nabla\phi\right] \tag{17}$$

$$\nabla x_W = \frac{x_W N_{OH^-} - x_{OH^-}N_W}{c_T D^{M,eff}_{W,OH^-}} - \frac{N_W}{c_T D^{M,eff}_{W,AEM}} - \frac{B_o}{\eta D^{M,eff}_{W,AEM}}\left[\nabla P + c_{OH^-}F\nabla\phi\right] \tag{18}$$

Transport by convection is especially important at high hydration levels in which porous structure of the membrane swells enough to give place to continuous channels that connect the anode with cathode, so an effective pressure gradient can be established. In addition, convection by electro-osmotic drag must be considered at high concentrations of hydroxide ions when electric potential gradients could be significant.

3.3. Migration

It is defined as the motion of charged species due to electric potential gradients resulting from the electrostatic interactions between them. Transport by migration can be described according to Ohm's law:

$$i = -\sigma\nabla\phi \tag{19}$$

in which σ is the ionic conductivity of the membrane, ϕ the electric potential and i the current density. The latter can be related to molar flux by the following definition:

$$i = F\sum_i^n z_i N_i \tag{20}$$

When either diffusion or convection takes place simultaneously with migration, the generalized Stefan-Maxwell equations (GSME) for a system of n-particles can be used [29]:

$$\frac{c_i}{RT}\nabla\tilde{\mu}_i = \sum_{j\neq i}^n \frac{x_i N_j - x_j N_i}{D^{eff}_{i,j}} - \frac{N_i}{D^{eff}_{i,AEM}} - \frac{B_o c_i}{\eta D^{eff}_{i,AEM}}\left[\nabla P + \left(\sum_i^n z_i c_i\right)F\nabla\phi\right] \tag{21}$$

which is analogous to Eq. (16) but expressed in terms of electrochemical potential gradients:

$$\nabla\tilde{\mu}_i = \nabla\mu_i + z_i F\nabla\phi \tag{22}$$

Alternatively, the dilute-solution or concentrate-solution approaches can be applied. In the dilute-solution approximation, interactions between solute molecules are neglected, and the Nernst-Planck equation can be used [32]:

$$N_i = -z_i u_i F c_i \nabla\phi - D_i \nabla c_i + c_i v_{conv} \tag{23}$$

in which u_i is the mobility of specie i. The terms of the right correspond respectively to migration, diffusion, and convection (in which electro-osmotic drag is neglected). In addition, the Nernst-Einstein equation can be used to relate ionic mobility (and conductivity) with diffusivity and reduce the number of transport properties in Eq. (23) [32]:

$$D_i = RTu_i \tag{24}$$

However, Eqs. (23) and (24) apply rigorously just at infinite dilution (<0.01 mol/dm^3 [49]), which most of the time is not the case for a membrane in an AEMFC. Therefore, concentrated-solution theory is preferred if enough information about the required parameters and transport properties is available [32]. Under this approach, the following set of equations for hydroxide ions and water can be obtained (see detailed derivation in Refs. [50, 51]):

$$i_{OH^-} = -\sigma \nabla \phi_{OH^-} - \frac{\sigma \xi}{F} \nabla \mu_w \tag{25}$$

$$N_w = -\frac{\sigma \xi}{F} \nabla \phi_{OH^-} - \left(\alpha + \frac{\sigma \xi^2}{F^2} \right) \nabla \mu_w \tag{26}$$

in which ξ is the electro-osmotic drag coefficient and α is a transport coefficient that can be related either to a hydraulic pressure gradient or a concentration gradient through the definition of chemical potential [32]:

$$\nabla \mu_w = RT \nabla \ln a_w + \overline{V}_w \nabla P \tag{27}$$

in which a_w is the activity and $\overline{V}_w$ the molar volume. Along with diffusion, transport by migration usually has a strong effect in the mobility of hydroxide ions even at low ionic concentrations. Therefore, application of concentrated-solution theory could be considered essential for the correct description of full transport phenomena of hydroxide ions. However, lack of information about most of the transport properties and parameters in the model for anion-exchange membranes limits greatly its use. Therefore, mathematical expressions for those properties in anion-exchange membranes are based on correlations fully characterized and validated for proton-exchange membranes [29], on the basis that transport phenomena in both systems should be equivalent as was mentioned at the beginning of this section.

4. Conclusions, challenges, and future work

Although PEMFCs are nowadays the main and most attractive fuel cell technology for mobile applications, AEMFCs have significant advantages over PEMFCs that give them a greater potential for massive production at low cost: (a) the oxygen reduction reaction is more favorable in alkaline medium. Therefore, catalysts as silver or nickel, which are less expensive than platinum (predominantly used in PEMFCs and their major drawback) can be used [52]. (b) Corrosion problems are reduced because metals and coal generally used in bipolar plates, current collectors and catalysts are chemically more stable in alkaline than in acid medium.

This allows the use of less expensive components such as current collectors of nickel and thin metal bipolar plates [8]. (c) The oxidation of alcohols such as methanol and ethanol is more favorable in alkaline medium. This allows a more efficient alcohol oxidation, and the amount of platinum and ruthenium in catalysts can be reduced [53]. However, more intensive research is required.

Homogeneous membranes are currently the most efficient electrolytes for AEMFCs because of their high ionic conductivity and durability in comparison with heterogeneous membranes and interpenetrated polymer networks. However, they have strong drawbacks that need to be overcome. Membranes produced by copolymerization using chloromethylstyrene and divinylbenzene as their base materials have shown both high ionic conductivities and chemical stability at fuel cell conditions, but as was mentioned in Section 2, they are impractical because of the low availability of chloromethylstyrene and high cost of divinylbenzene. Conductive polymers functionalized by radiation grafting procedures have been extensively researched and have the highest ionic conductivities among the listed membrane types. In fact, most of the commercial membranes nowadays belong to this category. However, as membranes produced by coplymerization, they are very expensive and unfeasible for a large-scale production. Finally, homogeneous membranes functionalized by chemical modification can be produced at low cost and be as conductive and efficient as membranes produced by copolymerization and radiation grafting. Therefore, they are the most promising materials for a future mass production of AEMFCs. The main problem with these materials is the attachment process of the cationic functional groups and their chemical stability. First, fabrication procedures need to be improved and standardized so they can be effectively implement at large-scale production. Second, membranes of this category use trimethylammonium-based chains as the cationic functional groups, which are very reactive with hydroxide ions. In consequence, they have poor stability at operational conditions of the fuel cell and tend to degrade at early operation times.

Theoretical understanding of the transport mechanisms taking place within anion-exchange membranes is a straightforward way to design effective procedures to fabricate anion-exchange membranes with the required characteristics to operate efficiently and stably at the conditions of an AEMFC. Unfortunately, there are a lot of conceptual gaps and lack of information that currently limits strongly this possibility. It is required a better understanding and particularization of the transport theories behind those models to anion-exchange membranes. For instance, great part of the transport properties and parameters required by the most rigorous models (like GSME and concentrated-solution theory) have not been experimentally measured or mathematically defined for anion-exchange membranes, so they still are calculated from correlations and theories extensively studied and validated for proton-exchange membranes regardless their application could not be correct for alkaline medium. Moreover, the characteristics of some of the transport mechanisms involved in the mobility of hydroxide ions are not exactly known. This is, for example, the case of the mechanism considered to have the major contribution to the hydroxide mobility through hydrated anion-exchange membranes, that is, Grotthuss mechanism. Mobility studies for hydroxide ions in pure aqueous medium have approached the basic characteristics of the mechanism, but it has not been done yet an extension of those studies to anion-exchange membranes to take into account key aspects that could affect dramatically its development such as: (a) water content of the conductive polymer, which affects its morphology and distribution of its cationic functional groups and (b) the type of cationic functional group, which is responsible of the electrostatic forces exerted on the hydroxide ions and water, steric and frictional effects due to the structure and size of the cationic group and the electrical double layer effects due to the surface distribution of electrostatically attracted molecules [14, 33, 34, 45].

All the above mentioned approaches are nowadays active research areas of high interest. If successful alternatives to improve the current limitations of the existent polymeric systems could be obtained from the co-development of experimental and theoretical studies, AEMFCs could not only exceed in efficiency and economy the PEMFCs, but also become a competitive, feasible, and sustainable technology to other alternative power generation sources especially for portable and stationary applications at low temperature.

Acknowledgements

Special acknowledgment to Universidad Nacional de Colombia-Sede Medellín and the Curricular Area of Chemical and Petroleum Engineering for the financial support for the development and publication of this work.

Author details

Sergio Castañeda Ramírez and Rafael Ribadeneira Paz*

*Address all correspondence to: reribade@unal.edu.co

Department of Processes and Energy, Facultad de Minas, Universidad Nacional de Colombia-Sede Medellín, Colombia

References

[1] Wang Y, Chen KS, Mishler J, Cho SC, Adroher XC. A review of polymer electrolyte membrane fuel cells: Technology, applications, and needs on fundamental research. Applied Energy. 2011;**88**(4):981-1007. DOI: 10.1016/j.apenergy.2010.09.030

[2] Das V, Padmanaban S, Venkitusamy K, Selvamuthukumaran R, Blaabjerg F, Siano P. Recent advances and challenges of fuel cell based power system architectures and control —A review. Renewable and Sustainable Energy Reviews. 2017;**73**:10-18. DOI: 10.1016/j.rser.2017.01.148

[3] Merle G, Wessling M, Nijmeijer K. Anion exchange membranes for alkaline fuel cells: A review. Journal of Membrane Science. 2011;**377**(1–2):1-35. DOI: 10.1016/j.memsci.2011.04.043

[4] Varcoe JR, Atanassov P, Dekel DR, Herring AM, Hickner MA, Kohl PA, et al. Anion-exchange membranes in electrochemical energy systems. Energy & Environmental Science. 2014;**7**:3135-3191. DOI: 10.1039/c4ee01303d

[5] Couture G, Alaaeddine A, Boschet F, Ameduri B. Polymeric materials as anion-exchange membranes for alkaline fuel cells. Progress in Polymer Science. 2011;**36**(11):1521-1557. DOI: 10.1016/j.progpolymsci.2011.04.004

[6] Antolini E, Gonzalez ER. Alkaline direct alcohol fuel cells. Journal of Power Sources. 2010; **195**(11):3431-3450. DOI: 10.1016/j.jpowsour.2009.11.145

[7] Matsuoka K, Iriyama Y, Abe T, Matsuoka M, Ogumi Z. Alkaline direct alcohol fuel cells using an anion exchange membrane. Journal of Power Sources. 2005;**150**:27-31. DOI: 10.1016/j.jpowsour.2005.02.020

[8] Varcoe JR, Slade RCT. Prospects for alkaline anion-exchange membranes in low temperature fuel cells. Fuel Cells. 2005;**5**(2):187-200. DOI: 10.1002/fuce.200400045

[9] Varcoe JR. Investigations of the ex situ ionic conductivities at 30 degrees C of metal-cation-free quaternary ammonium alkaline anion-exchange membranes in static atmospheres of different relative humidities. Physical Chemistry Chemical Physics. 2007;**9**:1479-1486. DOI: 10.1039/B615478F

[10] Cifrain M, Kordesch K. Advances, aging mechanism and lifetime in AFCs with circulating electrolytes. Journal of Power Sources. 2004;**127**(1–2):234-242. DOI: 10.1016/j.jpowsour.2003.09.019

[11] Sollogoub C, Guinault A, Bonnebat C, Bennjima M, Akrour L, Fauvarque JF, et al. Formation and characterization of crosslinked membranes for alkaline fuel cells. Journal of Membrane Science. 2009;**335**(1–2):37-42. DOI: 10.1016/j.memsci.2009.02.027

[12] Pivovar B. Alkaline Membrane Fuel Cell Workshop Final Report. In: Alkaline Membrane Fuel Cell Workshop; 8-9 May 2011; Arlington. U.S. Department of Energy, Office of Scientific and Technical Information; 2011. pp. 1-24

[13] Bidault F, Brett DJL, Middleton PH, Brandon NP. Review of gas diffusion cathodes for alkaline fuel cells. Journal of Power Sources. 2009;**187**(1):39-48. DOI: 10.1016/j.jpowsour.2008.10.106

[14] Cheng J, He G, Zhang F. A mini-review on anion exchange membranes for fuel cell applications: Stability issue and addressing strategies. International Journal of Hydrogen Energy. 2015;**40**(23):7348-7360. DOI: 10.1016/j.ijhydene.2015.04.040

[15] Wu Y, Wu C, Varcoe JR, Poynton SD, Xu T, Fu Y. Novel silica/poly(2,6-dimethyl-1,4-phenylene oxide) hybrid anion-exchange membranes for alkaline fuel cells: Effect of silica content and the single cell performance. Journal of Power Sources. 2010;**195**(10):3069-3076. DOI: 10.1016/j.jpowsour.2009.11.118

[16] Zeng QH, Liu QL, Broadwell I, Zhu AM, Xiong Y, Tu XP. Anion exchange membranes based on quaternized polystyrene-block-poly(ethylene-ran-butylene)-block-polystyrene for

direct methanol alkaline fuel cells. Journal of Membrane Science. 2010;**349**(1–2):237-243. DOI: 10.1016/j.memsci.2009.11.051

[17] Chempath S, Boncella JM, Pratt LR, Henson N, Pivovar BS. Density functional theory study of degradation of Tetraalkylammonium Hydroxides. Journal of Physical Chemistry C. 2010; **114**(27):11977-11983. DOI: 10.1021/jp9122198

[18] Long H, Kim K, Pivovar BS. Hydroxide degradation pathways for substituted trimethylammonium cations: A DFT study. Journal of Physical Chemistry C. 2012;**116**(17):9419-9426. DOI: 10.1021/jp3014964

[19] Ponce-González J, Whelligan DK, Wang L, Bance-Soualhi R, Wang Y, Peng Y, et al. High performance aliphatic-heterocyclic benzyl-quaternary ammonium radiation-grafted anion-exchange membranes. Energy & Environmental Science. 2016;**9**:3724-3735. DOI: 10.1039/C6EE01958G

[20] Maurya S, Shin S-H, Kim Y, Moon S-H. A review on recent developments of anion exchange membranes for fuel cells and redox flow batteries. RSC Advances. 2015;**5**(47): 37206-37230. DOI: 10.1039/c5ra04741b

[21] Li N, Yan T, Li Z, Thurn-Albrecht T, Binder WH. Comb-shaped polymers to enhance hydroxide transport in anion exchange membranes. Energy & Environmental Science. 2012;**5**(7):7888. DOI: 10.1039/c2ee22050d

[22] Shevchenko VV, Gumennaya MA. Synthesis and properties of anion-exchange membranes for fuel cells. Theoretical and Experimental Chemistry. 2010;**46**(3):139-152. DOI: 10.1007/s11237-010-9131-4

[23] Zhao TS, Li YS, Shen SY. Anion-exchange membrane direct ethanol fuel cells: Status and perspective. Frontiers of Energy and Power Engineering in China. 2010;**4**(4):443-458. DOI: 10.1007/s11708-010-0127-5

[24] Di Noto V, Zawodzinski TA, Herring AM, Giffin GA, Negro E, Lavina S. Polymer electrolytes for a hydrogen economy. International Journal of Hydrogen Energy. 2012;**37**(7):6120-6131. DOI: 10.1016/j.ijhydene.2012.01.080

[25] Fauvarque JF, Guinot S, Bouzir N, Salmon E, Penneau JF. Alkaline poly(ethylene oxide) solid polymer electrolytes. Application to nickel secondary batteries. Electrochimica Acta. 1995;**40**(13–14):2449-2453. DOI: 10.1016/0013-4686(95)00212-W

[26] Vassal N, Salmon E, Fauvarque JF. Electrochemical properties of an alkaline solid polymer electrolyte based on P(ECH-co-EO). Electrochimica Acta. 2000;**45**(8–9):1527-1532. DOI: 10.1016/S0013-4686(99)00369-2

[27] Guinot S, Salmon E, Penneau JF, Fauvarque JF. A new class of PEO-based SPEs: structure, conductivity and application to alkaline secondary batteries. Electrochimica Acta. 1998;**43**: 1163-1170. DOI: 10.1016/S0013-4686(97)10015-9

[28] Xu T. Ion exchange membranes: State of their development and perspective. Journal of Membrane Science. 2005;**263**(1–2):1-29. DOI: 10.1016/j.memsci.2005.05.002

[29] Grew KN, Chiu WKS. A Dusty Fluid model for predicting hydroxyl anion conductivity in alkaline anion exchange membranes. Journal of the Electrochemical Society. 2010;**157**(3): B327. DOI: 10.1149/1.3273200

[30] Feng L, Zhang X, Wang C, Li X, Zhao Y, Xie X, et al. Effect of different imidazole group positions on the hydroxyl ion conductivity. International Journal of Hydrogen Energy. 2016;**41**(36):16135-16141. DOI: 10.1016/j.ijhydene.2016.05.252

[31] Takaba H, Hisabe T, Shimizu T, Alam K. Molecular modeling of OH- transport in poly (arylene ether sulfone ketone)s containing quaternized ammonio-substituted fluorenyl groups as anion exchange membranes. Journal of Membrane Science. 2017;**522**:237-244. DOI: 10.1016/j.memsci.2016.09.019

[32] Weber AZ, Borup RL, Darling RM, Das PK, Dursch TJ, Gu W, et al. A critical review of modeling transport phenomena in polymer-electrolyte fuel cells. Journal of the Electrochemical Society. 2014;**161**(12):1254-1299. DOI: 10.1149/2.0751412jes

[33] Choi P, Jalani NH, Datta R. Thermodynamics and proton transport in Nafion II. Proton diffusion mechanisms and conductivity. Journal of the Electrochemical Society. 2005; **152**(3):E123. DOI: 10.1149/1.1859814

[34] Weber AZ, Newman J. Transport in polymer-electrolyte membranes I. Physical model. Journal of the Electrochemical Society. 2003;**150**(7):A1008. DOI: 10.1149/1.1580822

[35] Weber AZ, Newman J. Transport in polymer-electrolyte membranes III. Model validation in a simple fuel-cell model. Journal of the Electrochemical Society. 2004;**151**(2):A326. DOI: 10.1149/1.1639158

[36] Tuckerman ME, Marx D, Parrinello M. The nature and transport mechanism of hydrated hydroxide ions in aqueous solution. Nature. 2002;**417**:925-929. DOI: 10.1038/nature00794.1

[37] Tuckerman ME, Chandra A, Marx D. Structure and dynamics of OH-(aq). Accounts of Chemical Research. 2006;**39**(2):151-158. DOI: 10.1021/ar040207n

[38] Marx D, Chandra A, Tuckerman ME. Aqueous basic solutions: Hydroxide solvation, structural diffusion, and comparison to the hydrated proton. Chemical Reviews. 2010; **110**(4):2174-2216. DOI: 10.1021/cr900233f

[39] Atkins P, De Paula J. Atkins' Physical Chemistry. 8th ed. The Elements of Physical Chemistry. New York: Oxford University Press; 2006. p. 1085. DOI: 10.1039/c1cs15191f

[40] Ma Z, Tuckerman ME. On the connection between proton transport, structural diffusion, and reorientation of the hydrated hydroxide ion as a function of temperature. Chemical Physics Letters. 2011;**511**(4–6):177-182. DOI: 10.1016/j.cplett.2011.05.066

[41] Marx D. Throwing tetrahedral dice. Science. 2004;**303**:634-636. DOI: 10.1126/science.1094001

[42] Stillinger FH. Water revisited. Science. 1980;**209**(4455):451-457. DOI: 10.1126/science.209.4455.451

[43] Botti A, Bruni F, Imberti S, Ricci MA, Soper AK. Ions in water: The microscopic structure of concentrated NaOH solutions. The Journal of Chemical Physics. 2004;**120**(21):10154-10162. DOI: 10.1063/1.1705572

[44] McLain S, Imberti S, Soper A, Botti A, Bruni F, Ricci M. Structure of 2 molar NaOH in aqueous solution from neutron diffraction and empirical potential structure refinement. Physical Review B. 2006;**74**(9):94201. DOI: 10.1103/PhysRevB.74.094201

[45] Hwang GS, Kaviany M, Gostick JT, Kientiz B, Weber AZ, Kim MH. Role of water states on water uptake and proton transport in Nafion using molecular simulations and bimodal network. Polymer. 2011;**52**(12):2584-2593. DOI: 10.1016/j.polymer.2011.03.056

[46] Eikerling M, Kornyshev AA, Kuznetsov AM, Ulstrup J, Walbran S. Mechanisms of proton conductance in polymer electrolyte membranes. The Journal of Physical Chemistry. B. 2001;**105**(17):3646-3662. DOI: 10.1021/jp003182s

[47] Kreuer K. Fast proton conductivity: A phenomenon between the solid and the liquid state? Solid State Ionics. 1997;**94**(1–4):55-62. DOI: 10.1016/S0167-2738(96)00608-X

[48] Zhou X, Chen Z, Delgado F, Brenner D, Srivastava R. Atomistic simulation of conduction and diffusion processes in Nafion polymer electrolyte and experimental validation. Journal of the Electrochemical Society. 2007;**154**(1):B82. DOI: 10.1149/1.2388735

[49] Roušar I, Micka K, Kimla A. Electrochemical Engineering. Prague: Elsevier; 1986. p. 253

[50] Weber AZ, Newman J. Transport in polymer-electrolyte membranes II. Mathematical Model. Journal of the Electrochemical Society. 2004;**151**(2):A311. DOI: 10.1149/1.1639157

[51] Castañeda S, Sánchez CI. Modeling and analysis of ion transport through anion exchange membranes used in alkaline fuel cells. ECS Transactions. 2012;**50**(2):2091-2107. DOI: 10.1149/05002.2091ecst

[52] Hickner MA, Herring AM, Coughlin EB. Anion exchange membranes: Current status and moving forward. Journal of Polymer Science Part B: Polymer Physics. 2013;**51**(24):1727-1735

[53] Tripkovic AV, Popovic KD, Grgur BN, Blizanac B, Ross PN, Markovic NM. Methanol electrooxidation on supported Pt and PtRu catalysts in acid and alkaline solutions. Electrochimica Acta. 2002;**47**:3707-3714. DOI: 10.1002/polb.23395

Chapter 5

Ion Exchange in Geopolymers

José Ramón Gasca-Tirado,
Alejandro Manzano-Ramírez, Eric M. RiveraMuñoz,
Rodrigo Velázquez-Castillo, Miguel Apátiga-Castro,
Rufino Nava and Aarón Rodríguez-López

Additional information is available at the end of the chapter

http://dx.doi.org/10.5772/intechopen.80970

Abstract

Geopolymers have been widely used for construction and building materials. Nevertheless, some other applications have been found from their ability to be ion exchanged. An example is the encapsulation of heavy metals, but some others involve the ion exchange of the aluminosilicate structure to form photoactive particles or to link copper ions. In this chapter, we summarize some of the properties which make aluminosilicate inorganic polymer (geopolymers) ion exchangeable: the synthesized temperature, its effect over their porosity and their stoichiometric nature. Also, the effects of ion exchanging a geopolymer with an NH_4^+Cl, $(NH_4)_2TiO_2(C_2O_4)_2$ and $(CH_3)_4N^+Br$ are presented. The geopolymer was characterized by FT-IR, XRD, BET and MAS NMR, showing how a 100% of replacement was achieved for NH_4^+Cl. On the contrary, the efficiency was reduced in $(NH_4)_2TiO_2(C_2O_4)_2$ and $(CH_3)_4N^+Br$, effect ascribed to the fact of the molecular size that did not allow the counterions to reach the aluminum atoms in the geopolymer. Finally, the procedure followed to ion exchange a metakaolinite-based geopolymer is described, and the potential applications related are presented.

Keywords: geopolymer, ion exchange

1. Introduction

Since the origin of the term geopolymer to define those materials synthesized by the alkali activation of aluminosilicate monomers at ambient temperature, studies about geopolymers have been widely spread [1].

Examples include environmental alternatives for Portland cement [2], refractories [3] and ceramic precursor, for instance [4, 5].

Nevertheless, while these studies continue, others have explored new applications as immobilizations of waste, radioactive and toxic materials [6, 7], medicinal applications, catalytic materials [8, 9], catalytic materials for volatile organic degradation [9–11] and support material for optical applications like color holder, color pH indicator and photoluminescent material.

These new applications take advantage of the nonordered AlO_4-SiO_4 structure of geopolymers that results in a negative charged framework where alkali or alkaline earth cations (Na^+, K^+, Ca^{2+}, Li^+ and rarely Cs^+) need to be present to balance the charge of the geopolymer.

The chemical similarity of geopolymers to zeolites makes them an important former material to synthesize a wide variety of new ceramics through the ion exchange of the charge-balancing cation [8, 12].

To improve the studies and increase the amount of applications of geopolymers, it is necessary to understand clearly the core variables that determine a good ion exchange. Therefore, in this chapter, we review some of these core variables which make geopolymers ion exchangeable: the synthesis temperature was varied from 40 to 90°C to define a specific pore size distribution (PSD) within the samples, the removal of soluble species from the samples and the use of the right chemical compound according to the desired ion to be exchanged to avoid any change in the geopolymer structure after ion exchange. Finally, we describe the procedure followed to ion exchange a metakaolinite-based geopolymer and their potential applications.

2. Synthesis of geopolymers

Geopolymers are synthesized by the dissolution and polycondensation of an aluminosilicate powder in an alkaline silicate solution under hydrothermal conditions [13]. Even though there are different sources for solid aluminosilicate like slag, fly ash and mine tailing [2], the metakaolin is the preferred starting material chosen to synthesize geopolymers. This is mainly due to its high purity that results in a suitable geopolymer to be ion exchanged.

On the other hand, the available alkaline silicate solutions needed to synthesize a geopolymer are limited to solution in which cations are alkaline enough for synthesizing geopolymers (e.g. Na^+ or K^+) [14]. Therefore, the variety of geopolymers that can be synthesized is constrained to this issue.

The samples synthesized for this study were prepared by mechanically mixing stoichiometric amounts of metakaolin, sodium hydroxide, distilled water and sodium silicate (Na_2O/SiO_2 wt. ratio: 0.31) to follow the recommended molar ratios: SiO_2:Al_2O_3 = 2.89, NaO_2:SiO_2 = 0.29 and H_2O:Na_2O = 10.01 which gave geopolymers good mechanical properties (3.16, 0.34 and16.2 for the SiO_2/Al_2O_3, K_2O/SiO_2 and H_2O/K_2O, respectively, in the case of potassium geopolymers) [15, 16].

To have a homogeneous slurry, a mechanical mixing followed by 15 min of vibration was done. Final samples were casted into plastic containers, and they were left under constant temperature at 40 and 90°C for 24 h [16–19]. After cooling, geopolymers were cut with a diamond disc to have samples of around 1 g weight.

3. Geopolymer structure

As all geopolymers, the sample synthesis takes place in local coordination environment like zeolites. The silicate and aluminate units that form the geopolymer are completely disordered on length scales longer than 10 Å and are connected to form a disordered network

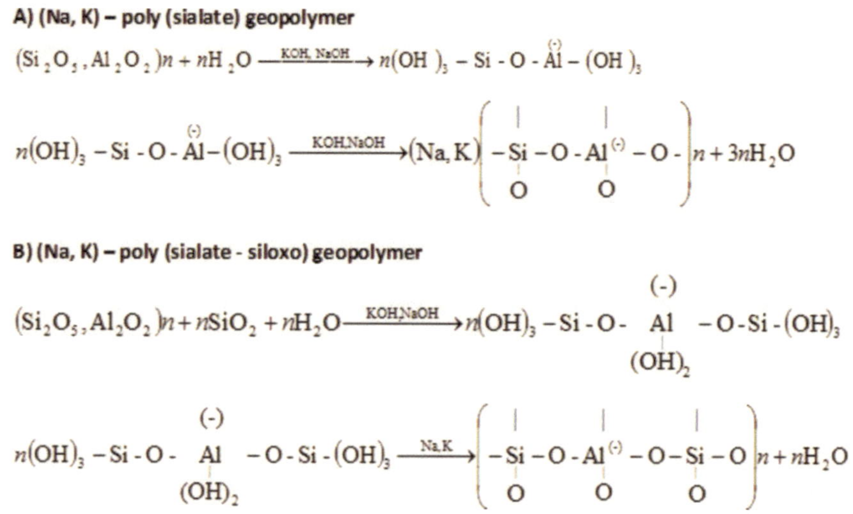

Figure 1. Formation of geopolymers: (A) poly(sialate) and (B) poly(sialate-siloxo).

Si:Al ratio	Potential applications	Si:Al ratio	Potential applications
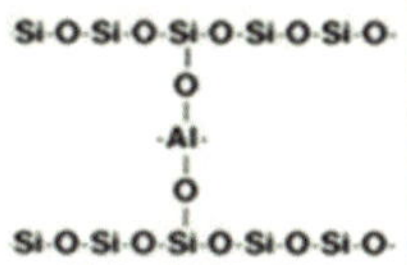 20:1 < Si:Al < 35:1	Fire resistant materials	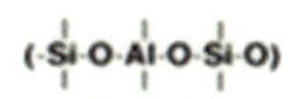 Si:Al 2:1	Cementitious material. Toxic waste encapsulation.
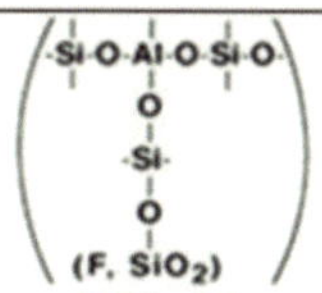 Si:Al >3:1	Thermal insulation material (200 to 600 °C). Tooling for SPF aluminium.	 Si:Al 1:1	Fire protection construction Materials. Building bricks.
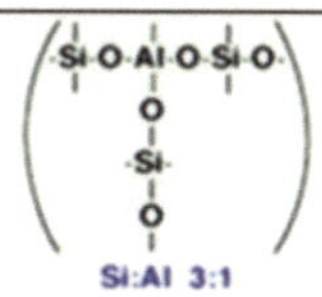 Si:Al 3:1	Thermal insulation material (200 a 1000 °C). Tooling for titanium processing.		

Figure 2. Potential applications of geopolymers according to the Si:Al ratio.

with the alkali cations in irregular sites [14]. These cations are present inside the geopolymer framework, and they are needed to balance the negative charge of Al_3^+ in $[AlO_4]^{-1}$ [20].

As it can be seen in **Figure 1** where geopolymers are classified according to the ratio of aluminates and silicates, the geopolymer synthesized was poly(sialate) geopolymers which are mainly used as cementitious and toxic waste encapsulation material (**Figure 2**).

According to their structure, the applications of geopolymer may be summarized in **Figure 2**.

4. Background of ion exchange

Ion exchange is a chemical reaction where an ion carrier material (ion exchanger) exchanges their own ions with others disposed in solution (counterions). Even though the ion exchanger is not dissolved, the ion exchanger must have an open network structure to let ions be dissolved in the aqueous solution [21].

The ion exchangers can be cationic, anionic or amphoteric in accordance with the charge of the balancing ion, and the number of ions exchanged should be equal to the proportion of free counterions in solution in agreement with the stoichiometric ratio of exchange.

In the case of geopolymers, which are X-ray amorphous materials formed by non-localized but well-distributed negative tetrahedral silicate and aluminate units [1, 15], they can be considered as cationic exchangers. Their negatively charged structure, balanced by alkali metals such as Na^+ or K^+ (**Figure 3**), can be completely hydrated and mobilized [3, 22]. This gives a lower bonding strength in comparison with zeolites [20] and an ease to be ion exchanged when they are in contact with solutions of chloride or nitrate of a desired cation.

The geopolymer structure is made up by silicon and aluminium atoms linked by oxygen (**Figure 3**), and it is not modified when the sodium atom, the exchangeable ion linked to the aluminum atom, is replaced with another cation. The maximum number of possible exchangeable cations, known as theoretical ion exchange capacity (TEC), is equal to the negative charge of the geopolymer network and can be measured by elemental analysis of aluminum. Nevertheless, it is not always possible to ion exchange the total amount of available exchangeable cations (TEC value), and a real exchange capacity (REC value) should be determined by an ion-exchange method [23, 24].

The general procedure of ion exchange can be formulated by Eq. (1). When the geopolymer, represented as Na^+G^- (ion exchanger), is ionized in an aqueous solution, the released cation (Na^+ or K^+) diffuses inside the material, transfers through the interphase boundary and finally arrives to the solution. If a salt MY is also dissolved in the solution (M^+ and Y^- ions), the Na^+ in geopolymer is replaced by an equivalent amount of counterion M^+ due to the electroneutrality requirement:

$$Na^+G^- + M^+Y^- \rightarrow M^+G^- + Na^+ + Y^- \tag{1}$$

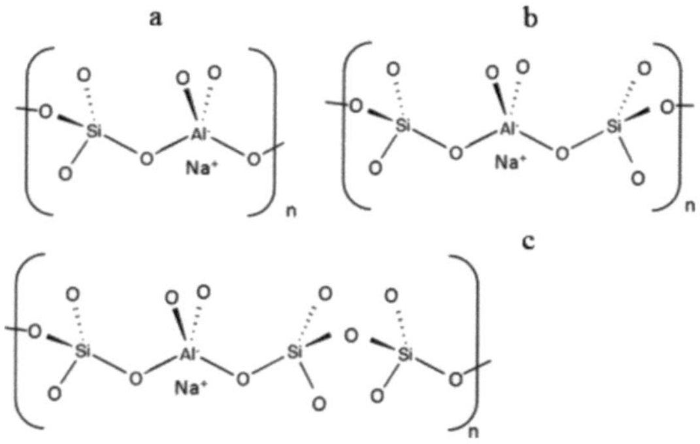

Figure 3. Position of the cation in the (a) sialate geopolymer, (b) sialate-siloxo geopolymer and (c) sialate-disiloxo geopolymer.

5. Procedure of ion exchange in geopolymers

Before an ion exchange is done, some soluble species from the samples must be removed. In this case, we followed a procedure described by Skorina [18]. The samples were rinsed several times with deionized water and a sequence of wash cycles, followed by a final treatment with a solution of KNO_3 until neutral "pH" was reached.

According to the desired ion to be exchanged with, the geopolymer was kept in contact for 12 h with solutions of NH_4Cl, $(NH_4)_2TiO_2(C_2O_4)_2$ and $(CH_3)_4NBr$ in concentration of 0.1 M with the aim of having a larger proportion of free counterions in solution and to guarantee a complete ion exchange. Finally, the samples were removed from the solution, rinsed several times with deionized water and dried for 1 week at ambient temperature.

6. Test performed over geopolymers

The pore diameter and surface area of samples were done under water desorpted samples (120° C for 24 h) by nitrogen adsorption on a NOVA 2000e Quantachrome instruments employing the Barrett–Joyner–Halenda method of analysis.

The geopolymer external surface was observed with a Philips XL-30 ESEM scanning electron microscope coupled with an EDAX energy-dispersive X-ray microanalyzer. The extent of ion exchange was determined by calculating the ratio of sodium with aluminum in the geopolymer as it was described by O'Connor [14].

The modification to the geopolymer structure was followed with a RIGAKU Ultima IV X-ray diffractometer using Cu irradiation (λ = 1.5406 Å), and the chemical composition was analyzed with a Perkin Elmer Spectrum Two using the IR-KBr technique.

Finally, ^{27}Al and ^{29}Si MAS NMR was performed to distinguish the atoms surrounding the aluminum atom with a Bruker Avance III HD 400 MHZ MAS NMR. This equipment was operated at 15 KHz for ^{27}Al and 8 KHz for ^{29}Si.

7. Effect of ion exchange over geopolymers

The total surface area and pore diameter of geopolymers synthesized at 40 and 90°C are summarized in **Table 1**. As it can be observed, the main factor that modified these parameters was the temperature synthesis. In addition, a reduction in pore width and the increase in specific surface were observed in all samples when the geopolymer was ion exchanged.

Modifications on the geopolymer surface were also observed in **Figure 4**. The ion exchanged samples were settled of small particles that occupied the void sites of the former geopolymer but leaving unchanged the geopolymer structure.

Counterion	Geopolymers at 40°C				Geopolymers at 90°C			
	Before ion exchange		After ion exchange		Before ion exchange		After ion exchange	
	S ($m^2 g^{-1}$)	D (nm)	S ($m^2 g^{-1}$)	D (nm)	S ($m^2 g^{-1}$)	D (nm)	S ($m^2 g^{-1}$)	D (nm)
NH_4Cl	45.88	19.56	123.84	2.19	27.21	30.52	491.33	2.45
$(NH_4)_2TiO_2(C_2O_4)_2$	45.88	19.56	526.54	2.19	27.21	30.52	195.99	2.19
$(CH_3)_4NBr$	45.88	19.56	150.14	2.19	27.21	30.52	450.17	2.7

Table 1. Surface area and pore diameter of geopolymers at 40 and 90°C, before and after ion exchange.

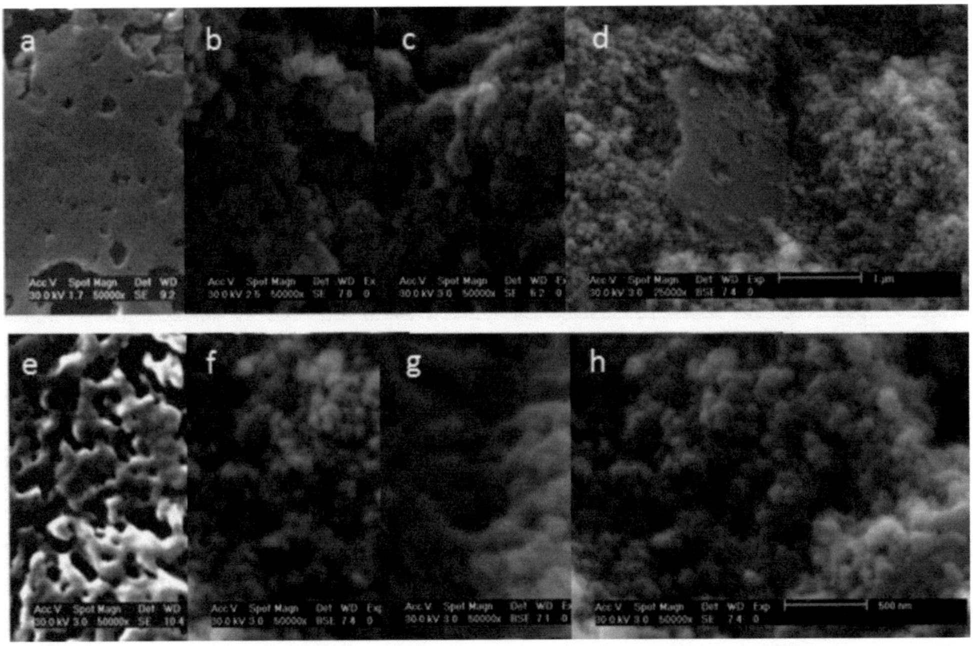

Figure 4. Effect of the ion exchange over the geopolymer surface. (a) Geopolymer at 40°C, (b) geopolymer at 40°C ion exchanged with NH_4Cl, (c) geopolymer at 40°C ion exchanged with $(NH_4)_2TiO_2(C_2O_4)_2$, (d) geopolymer at 40°C ion exchanged with $(CH_3)_4NBr$, (e) geopolymer at 90°C, (f) geopolymer at 90°C ion exchanged with NH_4Cl, (g) geopolymer at 90°C ion exchanged with $(NH_4)_2TiO_2(C_2O_4)_2$ and (h) geopolymer at 90°C ion exchanged with $(CH_3)_4NBr$.

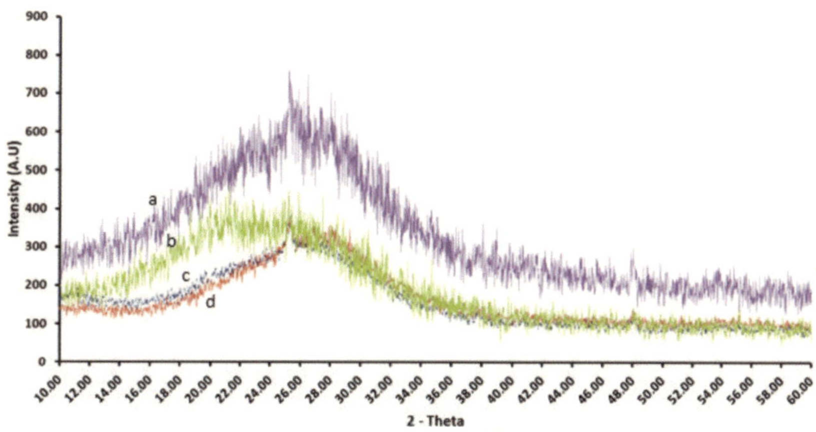

Figure 5. XRD pattern of (a) former geopolymer, (b) geopolymer ion exchanged with $(NH_4)_2TiO_2(C_2O_4)_2$, (c) geopolymer ion exchanged with NH_4Cl and (d) geopolymer ion exchanged with $(CH_3)_4NBr$.

This observation was supported by results obtained by XRD (**Figure 5**), where the amorphous structure of the geopolymer, represented by a hump at 2Θ = 28°, remained after ion exchange.

On the other hand, the FT-IR results of geopolymers at 40°C and 90°C before and after being ion exchanged are shown in **Figures 6** and **7**. As it can be observed in all spectra, the vibration modes corresponding to amorphous sodium aluminosilicate (1645.44, 878.56 cm^{-1}), symmetric stretching of Si-O-Al, Si-O-Si (706.9 cm^{-1}) and flexion of Si-O-Si and O-Si-O (571.73 cm^{-1})

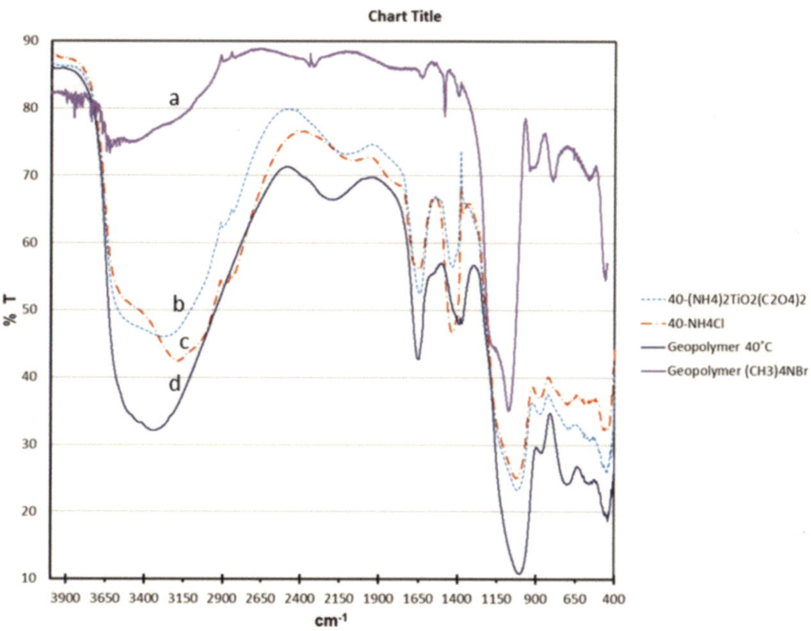

Figure 6. FT-IR of (a) geopolymer at 40°C ion exchanged with $(CH_3)_4NBr$, (b) geopolymer at 40°C ion exchanged with $(NH_4)_2TiO_2(C_2O_4)_2$, (c) geopolymer at 40°C ion exchanged with NH_4Cl and (d) geopolymer at 40°C.

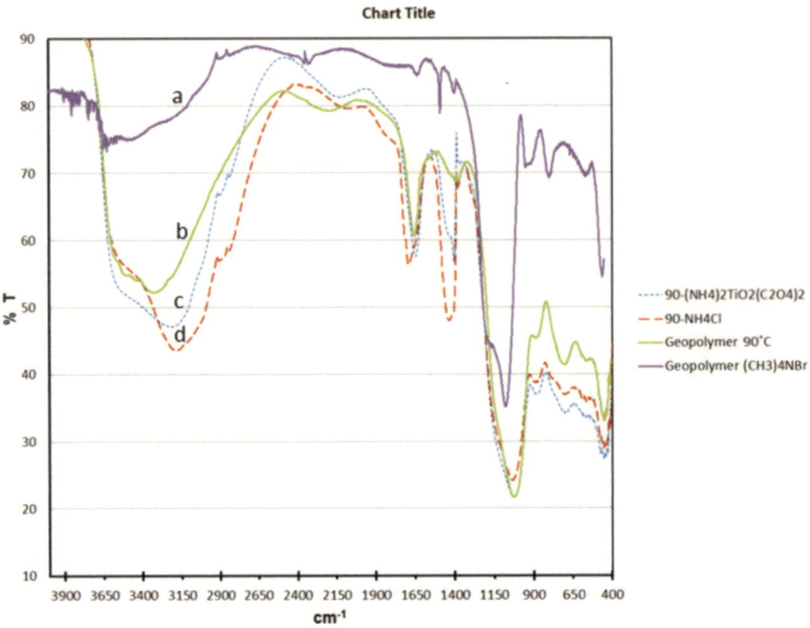

Figure 7. FT-IR of (a) geopolymer at 90°C ion exchanged with $(CH_3)_4NBr$ (b) geopolymer at 90°C, (c) geopolymer at 90°C ion exchanged with $(NH_4)_2TiO_2(C_2O_4)_2$ and (d) geopolymer at 90°C ion exchanged with NH_4Cl.

[25, 26] remained as a former geopolymer. The only modifications observed between spectra were those ascribed to vibration modes representative of their counterions.

Whenever a geopolymer is synthesized, the three-coordinated aluminum in metakaolin, Al (IV), Al (V) and Al (VI) (chemical shift: 49–80, 35–40 and −5–15 ppm, respectively), are transformed into Al (IV) to form the geopolymer structure [27]. As it can be observed in **Figure 8**, this structure remained in all ion exchanged samples, and in some occasions, Al (V) and AI (VI) arose at 30 ppm and 2 ppm as in the case of $(CH_3)_4NBr$ because of the size of the counterion mainly.

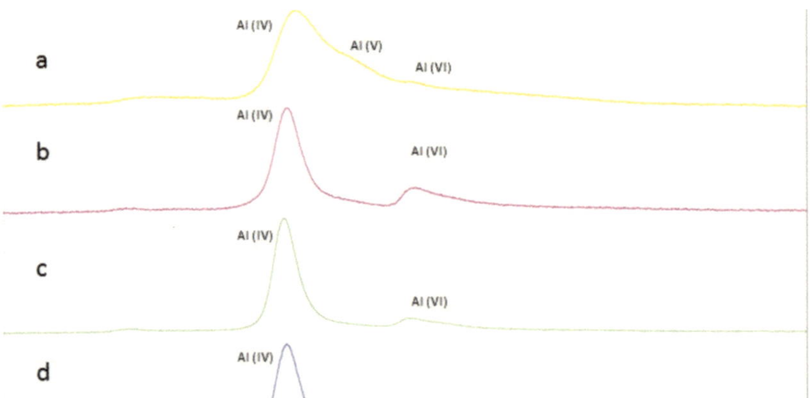

Figure 8. ^{27}Al MAS NMR of (a) geopolymer ion exchanged with $(CH_3)_4NBr$, (b) geopolymer ion exchanged with NH_4Cl, (c) geopolymer ion exchanged with $(NH_4)_2TiO_2(C_2O_4)_2$ and (d) former geopolymer.

http://dx.doi.org/10.5772/intechopen.80970

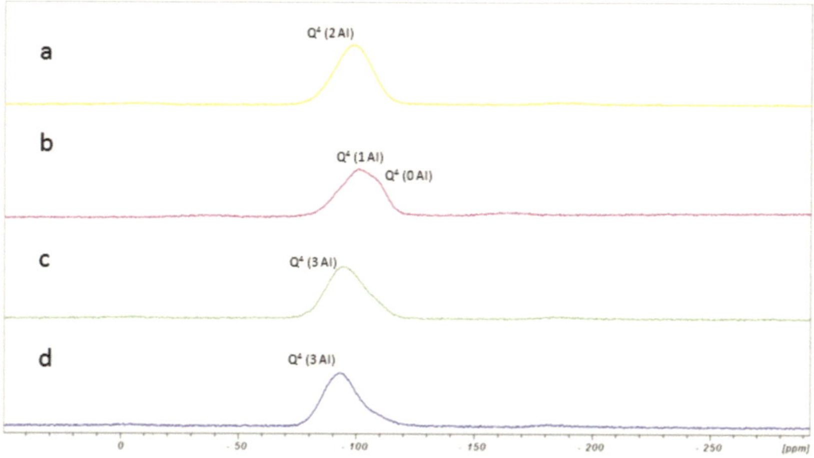

Figure 9. ^{29}Si MAS NMR of (a) geopolymer ion exchanged with $(CH_3)_4NBr$, (b) geopolymer ion exchanged with NH_4Cl, (c) geopolymer ion exchanged with $(NH_4)_2TiO_2(C_2O_4)_2$ and (d) former geopolymer.

Counterion	Real exchange capacity (%)
NH_4^+	100
$(NH_4)_2TiO_2(C_2O_4)_2$	80
$(CH_3)_4N^+$	85
Li^+	82
Cd^+	78
Mg^{2+}	57

Table 2. Real exchange capacity of geopolymers.

In the case of the ^{29}Si MAS NMR, all the geopolymer spectra presented a resonance peak between −90 and −100 ppm which were ascribed to tetrahedral silicon surrounded by two aluminum atoms [28] (**Figure 9**).

Finally, the real exchange capacity of samples is summarized in **Table 2**. A total replacement of sodium cation by NH_4^+ and a reduction on samples ion exchanged with $(NH_4)_2TiO_2(C_2O_4)_2$ and $(CH_3)_4N^+$ were determined. Results reported by O'Connor for Li^+, Cd^+ or Mg^{2+} are summarized, as a comparative, in **Table 2** [14].

8. Conclusions

In this chapter, the effects of ion exchanging a geopolymer with NH_4^+Cl, $(NH_4)_2TiO_2(C_2O_4)_2$ and $(CH_3)_4N^+Br$ were analyzed. Results were compared with the ion exchange with Li^+, Cd^+ and Mg^{2+} reported by O'Connor. A 100% of replacement was achieved for NH_4^+Cl. Nevertheless, the efficiency was reduced in $(NH_4)_2TiO_2(C_2O_4)_2$ and $(CH_3)_4N^+Br$ probably due to their molecular size that inhibited the counterions to reach the aluminum atom.

As it was observed by FT-IR, the ion exchange procedure did not modify the amorphous structure of geopolymers and the vibration modes ascribed to the symmetric stretching of Si-O-Al and Si-O-S, and the flexion energy of Si-O-Si and O-Si-O remained unmodified after the process. This can be explained if it is considered that the exchanged atom, Na^+, does not take part of the rigid structure of the geopolymer.

A similar conclusion was obtained from the XRD pattern, where the characteristic hump of amorphous geopolymer remained in all samples. Same results were observed in geopolymers synthesized at 40 and 90°C. The only observable change in this case was the superficial morphology which was measured by BET. Results gave a more porous material for samples synthesized at 40°C, but the extent of ion exchange was not affected by this parameter as it was summarized in **Table 2**.

In addition, the ^{27}Al and ^{29}Si MAS NMR spectra exposed the presence of Al (IV) and tetrahedral silicon, respectively, as an indicative of a structural order. In some cases, Al (V) and Al (VI) arose as it was reported by O'Connor for Mg^{2+}, Pb^{2+} and Cd^{2+} ions.

Finally, because the variety of geopolymers that can be produced by direct synthesis is limited to the selection of Na^+ or K^+ cations, the aim to find new application for geopolymer should lead research to find more exchange counterions.

Author details

José Ramón Gasca-Tirado[1]*, Alejandro Manzano-Ramírez[2], Eric M. RiveraMuñoz[3], Rodrigo Velázquez-Castillo[4], Miguel Apátiga-Castro[3], Rufino Nava[4] and Aarón Rodríguez-López[5]

*Address all correspondence to: ragatsi99@yahoo.com

1 Universidad de Guanajuato, Campus Celaya-Salvatierra, Celaya, Guanajuato, México

2 Centro de Investigaciones y Estudios Avanzados del I.P.N., Unidad Querétaro, Querétaro, México

3 Centro de Física Aplicada y Tecnología Avanzada, Universidad Nacional Autónoma de México, Querétaro, México

4 División de Investigación y Posgrado, Facultad de Ingeniería, Universidad Autónoma de Querétaro, Querétaro, México

5 Universidad Politécnica de Santa Rosa Jáuregui, Santa Rosa Jáuregui, Querétaro, México

References

[1] Davidovits J. Geopolymers. Journal of Thermal Analysis. 1991;**37**:1633-1656. DOI: 10.1007/BF01912193

[2] Duxson P, Provis JL, Lukey GC, van Deventer JSJ. The role of inorganic polymer technology in the development of 'green concrete'. Cement and Concrete Research. 2007;**37**:1590-1597. DOI: 10.1016/j.cemconres.2007.08.018

[3] Barbosa VFF, MacKenzie KJD. Synthesis and thermal behaviour of potassium sialate geopolymers. Materials Letters. 2003;**57**:1477-1482. DOI: 10.1016/S0167-577X(02)01009-1

[4] Kyoichi O, Yoshinori N, Swaroop V. Optical spectra of nickel-bearing silicate gels prepared by the geopolymer technique, with special reference to the low-temperature formation of liebenbergite (Ni_2SiO_4). Journal of the American Ceramic Society. 2004;**84**: 1717-1720. DOI: 10.1111/j.1151-2916.2001.tb00904.x

[5] Bell JL, Driemeyer PE, Kriven WM. Formation of ceramics from metakaolin-based geopolymers. Part II: K-based geopolymer. Journal of the American Ceramic Society. 2009;**92**:607-615. DOI: 10.1111/j.1551-2916.2008.02922.x

[6] Van Jaarsveld JGS, Van Deventer JSJ, Schwartzman A. The potential use of geopolymeric materials to immobilise toxic metals: Part II. Material and leaching characteristics. Minerals Engineering. 1999;**12**:75-91. DOI: 10.1016/S0892-6875(98)00121-6

[7] Khalil MY, Merz E. Immobilization of intermediate-level wastes in geopolymers. Journal of Nuclear Materials. 1994;**211**:141-148. DOI: 10.1016/0022-3115(94)90364-6

[8] O'Connor SJ, MacKenzie KJD. Synthesis, characterisation and thermal behaviour of lithium aluminosilicate inorganic polymers. Journal of Materials Science. 2010;**45**: 3707-3713. DOI: 10.1007/s10853-010-4383-x

[9] Sazama P, Bortnovsky O, Dědeček J, Tvarůžková Z, Sobalík Z. Geopolymer based catalysts—New group of catalytic materials. Catalysis Today. 2011;**164**:92-99. DOI: 10.1016/j.cattod.2010.09.008

[10] Gasca-Tirado JR, Manzano-Ramírez A, Reyes-Araiza JL. The Potential Use of Geopolymer for Cleaning Air. 2017. DOI: 10.1016/B978-0-08-102001-2.00009-7

[11] Gasca-Tirado JR, Manzano-Ramírez A, Villaseñor-Mora C, Muñiz-Villarreal MS, Zaldivar-Cadena AA, Rubio-Ávalos JC, et al. Incorporation of photoactive TiO_2 in an aluminosilicate inorganic polymer by ion exchange. Microporous and Mesoporous Materials. 2012;**153**. DOI: 10.1016/j.micromeso.2011.11.026

[12] Oleg B, Jiri D, Zdenka T, Zdeněk S, Jan Š. Metal ions as probes for characterization of geopolymer materials. Journal of the American Ceramic Society. 2008;**91**:3052-3057. DOI: 10.1111/j.1551-2916.2008.02577.x

[13] Cengiz B, Gregory PK, Kevin CS, Waltraud MK. Synthesis and characterization of silicon carbide powders converted from metakaolin-based geopolymer. Journal of the American Ceramic Society. 2016;**99**:2521-2530. DOI: 10.1111/jace.14254

[14] O'Connor SJ, MacKenzie KJD, Smith ME, Hanna JV. Ion exchange in the charge-balancing sites of aluminosilicate inorganic polymers. Journal of Materials Chemistry. 2010;**20**:10234-10240. DOI: 10.1039/c0jm01254h

[15] Barbosa VFF, MacKenzie KJD, Thaumaturgo C. Synthesis and characterisation of materials based on inorganic polymers of alumina and silica: Sodium polysialate polymers. International Journal of Inorganic Materials. 2000;**2**:309-317. DOI: 10.1016/S1466-6049(00)00041-6

[16] Rogers JJ, MacKenzie KJD, Trompetter WJ. New phosphors synthesised by ion exchange of a metakaolin-based geopolymer. Applied Clay Science. 2018;**157**:1-7. DOI: 10.1016/j.clay.2018.02.025

[17] Gasca-Tirado JR, Manzano-Ramírez A, Vazquez-Landaverde PA, Herrera-Díaz EI, Rodríguez-Ugarte ME, Rubio-Ávalos JC, et al. Ion-exchanged geopolymer for photocatalytic degradation of a volatile organic compound. Materials Letters. 2014. DOI: 10.1016/j.matlet.2014.07.090

[18] Skorina T. Ion exchange in amorphous alkali-activated aluminosilicates: Potassium based geopolymers. Applied Clay Science. 2014;**87**:205-211. DOI: 10.1016/j.clay.2013.11.003

[19] Singhal A, Gangwar BP, Gayathry JM. CTAB modified large surface area nanoporous geopolymer with high adsorption capacity for copper ion removal. Applied Clay Science. 2017;**150**:106-114. DOI: 10.1016/j.clay.2017.09.013

[20] He P, Fu S, Yuan J, Rao J, Xu J, Wang P, et al. Celsian formation from barium-exchanged geopolymer precursor: Thermal evolution. Journal of the European Ceramic Society. 2017;**37**:4179-4185. DOI: 10.1016/j.jeurceramsoc.2017.04.074

[21] Kumar S, Jain S. History, introduction, and kinetics of ion exchange materials. Journal of Chemistry. 2013;**2013**:13. DOI: 10.1155/2013/957647

[22] van Deventer JSJ, Provis JL, Duxson P, Lukey GC. Reaction mechanisms in the geopolymeric conversion of inorganic waste to useful products. Journal of Hazardous Materials. 2007;**139**:506-513. DOI: 10.1016/j.jhazmat.2006.02.044

[23] Inglezakis VJ. The concept of "capacity" in zeolite ion-exchange systems. Journal of Colloid and Interface Science. 2005;**281**:68-79. DOI: 10.1016/j.jcis.2004.08.082

[24] Lehto J, Harjula R. Experimentation in ion exchange studies—The problem of getting reliable and comparable results. Reactive and Functional Polymers. 1995;**27**:121-146. DOI: 10.1016/1381-5148(95)00038-H

[25] Rees CA, Provis JL, Lukey GC, van Deventer JSJ. Attenuated total reflectance Fourier transform infrared analysis of fly ash geopolymer gel aging. Langmuir. 2007;**23**:8170-8179. DOI: 10.1021/la700713g

[26] Serra J, González P, Liste S, Chiussi S, León B, Pérez-Amor M, et al. Influence of the non-bridging oxygen groups on the bioactivity of silicate glasses. Journal of Materials Science. Materials in Medicine. 2002;**13**:1221-1225. DOI: 10.1023/A:1021174912802

[27] Škvára F, Kopecký L, Šmilauer V, Bittnar Z. Material and structural characterization of alkali activated low-calcium brown coal fly ash. Journal of Hazardous Materials. 2009;**168**:711-720. DOI: 10.1016/j.jhazmat.2009.02.089

[28] Lee SK, Stebbins JF. The degree of aluminum avoidance in aluminosilicate glasses. American Mineralogist. 1999;**84**(5-6):937-945. DOI: 10.2138/am-1999-5-631